Nuclear Proliferation:
An Annotated Biography

NUCLEAR PROLIFERATION:
AN ANNOTATED BIOGRAPHY

A.M. BABKINA

Nova Science Publishers, Inc.
Commack, New York

Editorial Production: Susan Boriotti
Office Manager: Annette Hellinger
Graphics: Frank Grucci and John T'Lustachowski
Information Editor: Tatiana Shohov
Book Production: Donna Dennis, Patrick Davin, Christine Mathosian
and Tammy Sauter
Circulation: Maryanne Schmidt
Marketing/Sales: Cathy DeGregory

Library of Congress Cataloging-in-Publication Data available upon request

ISBN 1-56072-646-6

Copyright © 1999 by Nova Science Publishers, Inc.
6080 Jericho Turnpike, Suite 207
Commack, New York 11725
Tele. 516-499-3103   Fax 516-499-3146
e-mail: Novascience@earthlink.net
e-mail: Novascil@aol.com
Web Site: http://www.nexusworld.com/nova

All rights reserved. No part of this book may be reproduced, stored in a retrieval system or transmitted in any form or by any means: electronic, electrostatic, magnetic, tape, mechanical photocopying, recording or otherwise without permission from the publishers.

The authors and publisher have taken care in preparation of this book, but make no expressed or implied warranty of any kind and assume no responsibility for any errors or omissions. No liability is assumed for incidental or consequential damages in connection with or arising out of information contained in this book.

This publication is designed to provide accurate and authoritative information with regard to the subject matter covered herein. It is sold with the clear understanding that the publisher is not engaged in rendering legal or any other professional services. If legal or any other expert assistance is required, the services of a competent person should be sought. FROM A DECLARATION OF PARTICIPANTS JOINTLY ADOPTED BY A COMMITTEE OF THE AMERICAN BAR ASSOCIATION AND A COMMITTEE OF PUBLISHERS.

Printed in the United States of America

Contents

Part 1 Books
    1985-Present (1999)   1

Part 2 Journals
    1994-Present (1999)   117

Subject Index   227

Author Index   235

Part 1
Books and Reports
1985-Present (1999)

1995, a new beginning for the NPT? /
1995, a new beginning for the NPT? / edited by Joseph F.
Pilat and Robert E. Pendley ; foreword by Hans Blix. -
- New York : Plenum Press, c1995.
xix, 309 p. : 24 cm. -- (Issues in international
security)
Includes bibliographical references and index.
ISBN 0-306-45001-1
1. Nuclear nonproliferation. 2. Treaty on the Non
-proliferation of Nuclear Weapons (1968)   I. Pilat,
Joseph F. II. Pendley, Robert E. III. Series.
JX1974.73.A15 1995
go7.1'74--dc20   95-11394

Against proliferation--towards general disarmament /
Against proliferation--towards general disarmament /
Wolfgang Liebert, Jürgen Scheffran (eds.). -- Münster
: Agenda Verlag, 1995.
238 p. : ill. ; 23 cm. -- (Agenda Frieden ; 9)
Includes bibliographical references (p. [232]-238).
ISBN 3-929440-27-X
1. Nuclear nonproliferation. 2. Nuclear
disarmament. I. Liebert, Wolfgang. II. Scheffran,
Jürgen. III. Series.
JX1974.73.A45 1995
341.7'33--dc20   95-206398

Agreement between the United States and the People's
Republic of China concerning peaceful uses of nuclear
energy :
Agreement between the United States and the People's
Republic of China concerning peaceful uses of nuclear
energy : message from the President of the United
States transmitting an agreement for cooperation
between the government of the United States of America
and the government of the People's Republic of China
concerning peaceful uses of nuclear energy ... to the
Committee on Foreign Affairs, U.S. House of
Representatives. -- Washington : U.S. G.P.O., 1985.
vi, 56 p. ; 24 cm.
At head of title: 99th Congress, 1st session.
Committee print.
Distributed to some depository libraries in
microfiche.
Shipping list no.: 85-934-P.
"October 1985."
Item 1017-A, 1017-B (microfiche)
Supt. of Docs. no.: Y 4.F 76/1:Un 35/74
1. Nuclear nonproliferation. 2. Nuclear energy--Law
and legislation--United States. 3. Nuclear energy-
-Law and legislation--China. I. United States.
President (1981-1989 : Reagan) II. United States.
Congress. House. Committee on Foreign Affairs.

Agreement between the United States and the ... (Cont'd)
Agreement between the United States and the ... (Cont'd)
JX1974.73.A47   1985
327.1'74--dc19
DGPO/DLC
for Library of Congress   86-600847
                                                                r89

Alves, Dora.
Alves, Dora.
   Anti-nuclear attitudes in New Zealand and Australia
/ Dora Alves. -- Washington, DC : National Defense
University Press : For sale by the Supt. of Docs.,
U.S. G.P.O., 1985.
   xv, 91 p. : ill., 2 maps ; 21 cm. -- (A national
security affairs monograph)
   Shipping list no.: 86-97-P.
   "December 1985"--Verso t.p.
   Includes bibliographical references.
   S/N 008-020-01053-7
   Item 378-H
   Supt. of Docs. no.: D 5.409:N 88
   $2.25
   1. National security--New Zealand. 2. National
security--Australia. 3. Nuclear warfare. 4. Nuclear
nonproliferation. 5. Antinuclear movement--New
Zealand. 6. Antinuclear movement--Australia.
I. National Defense University Press. II. Title.
III. Series.
UA874.3.A48   1985
355'.0330931--dc19
DGPO/DLC
for Library of Congress   86-601148
                                                                r92

The American atom :
   The American atom : a documentary history of nuclear
policies from the discovery of fission to the present
/ Philip L. Cantelon, Richard G. Hewlett, and Robert
C. Williams, editors. -- 2nd ed. -- Philadelphia :
University of Pennsylvania Press, 1991.
   xviii, 369 p. ; 24 cm.
   Includes index.
   ISBN 0-8122-3096-5 (cloth). -- ISBN 0-8122-1354-8
(paper)
   1. United States--Military policy. 2. Nuclear
weapons--United States--History--Sources. 3. Nuclear
energy--United States--History--Sources. 4. Nuclear
nonproliferation--History--Sources. 5. Deterrence
(Strategy)--History--20th century--Sources.
I. Cantelon, Philip L. (Philip Louis), 1940-   .
II. Hewlett, Richard G. III. Williams, Robert
Chadwell, 1938-
UA23.A597   1991
355.02'17'0973--dc20   91-31676

Arms control :
Arms control : alliances, arms sales, and the future / edited by Kenneth W. Thompson. -- Lanham, Md. : University Press of America ; [Charlottesville] : Miller Center, University of Virginia, c1993.
    xiii, 175 p. ; 24 cm. -- (W. Alton Jones Foundation series on arms control ; v. 19)
    ISBN 0-8191-8936-7 (cloth : alk. paper). -- ISBN 0-8191-8937-5 (pbk. : alk. paper)
    1. Nuclear arms control. 2. Nuclear nonproliferation. I. Thompson, Kenneth W., 1921- II. White Burkett Miller Center. III. Series.
JX1974.A6743 1993
327.1'74--dc20   92-34823

Arms control in a multi-polar world /
Arms control in a multi-polar world / James Brown, editor. -- Amsterdam : VU University Press, 1996.
    xxix, 267 p. ; 24 cm.
    Includes bibliographical references.
    ISBN 9053834990
    1. Arms control. 2. Nuclear arms control. 3. Nuclear nonproliferation. I. Brown, James, 1934 May 1-
KZ5624.A75 1996
341.7'33--dc21   97-111572

Arms on the market :
Arms on the market : reducing the risk of proliferation in the former Soviet Union / edited by Gary K. Bertsch and Suzette Grillot. -- New York : Routledge, 1998.
    p. cm.
    Includes bibliographical references (p.   ).
    ISBN 0-415-92058-2. -- ISBN 0-415-92059-0
    1. Nuclear industry--Former Soviet republics. 2. Nuclear industry--Security measures--Former Soviet republics. 3. Nuclear industry--Former Soviet republics--Safety measures. 4. Nuclear industry--Safety measures--International cooperation. 5. Nuclear nonproliferation. I. Bertsch, Gary K. II. Grillot, Suzette.
HD9698.F62A75 1998
327.1'74--dc21   97-47647
                                                                               CIP 5/98

Arms proliferation policy :
Arms proliferation policy : support to the Presidential Advisory Board / Marcy Agmon ... [et al.]. -- Santa Monica, CA : Rand, 1996.
    xxvi, 132 p. : ill ; 28 cm.
    "Prepared for the Office of the Secretary of Defense."
    "National Defense Research Institute."
    Research performed for the project, "RAND support for the President's Advisory Board on Arms

Arms proliferation policy : (Cont'd)
Arms proliferation policy : support to the ... (Cont'd)
Proliferation Policy"--P. iii.
Includes bibliographical references (p. 121-132).
"MR-771-OSD"--P. 4 of cover.
ISBN 0-8330-2403-5
1. Nuclear nonproliferation. 2. Export controls- -United States. 3. Technology transfer--government policy--United States. I. Agmon, Marcy, 1949- . II. United States. Presidential Advisory Board on Arms Proliferation Policy. III. United States. Dept. of Defense. Office of the Secretary of Defense. IV. National Defense Research Institute (U.S.) V. Rand Corporation.
JX1974.73.A767 1996   96-177877

Aronson, Shlomo, 1936-
Aronson, Shlomo, 1936-
 The Politics and strategy of nuclear weapons in the Middle East : opacity, theory, and reality, 1960-1991 : an Israeli perspective / Shlomo Aronson with the assistance of Oded Brosh. -- Albany : State University of New York Press, c1992.
 xiii, 398 p. ; 24 cm. -- (SUNY series in Israeli studies)
 Includes bibliographical references (p. 357-369) and indexes.
 ISBN 0-7914-1207-5 (cloth). -- ISBN 0-7914-1208-3 (paper)
 1. Middle East--Military policy. 2. Israel- -Military policy. 3. Nuclear weapons--Middle East. 4. Nuclear weapons--Israel. 5. Nuclear nonproliferation. I. Brosh, Oded. II. Title. III. Series.
UA832.A77 1992
355'.033056--dc20   91-46244

Arquilla, John.
Arquilla, John.
 Modeling decisionmaking of potential proliferators as part of developing counterproliferation strategies / John Arquilla, Paul K. Davis. -- Santa Monica, CA : Rand, 1994.
 xv, 37 p. : ill. ; 28 cm.
 "National Security Research Division."
 "Prepared for the Office of Research and Development, Central Intelligence Agency."
 Includes bibliographical references (p. 35-37).
 "MR-467."
 ISBN 0-8330-1577-X
 1. Nuclear nonproliferation. 2. Nuclear arms control--United States--Decision making. 3. Security, International. I. Davis, Paul K., 1943- . II. United States. Central Intelligence Agency. Office of Research and Development. III. Rand Corporation. National Security Research Division. IV. Title.

Arquilla, John. (Cont'd)
Arquilla, John. (Cont'd)
JX1974.73.A77 1994 95-112811
r96

At the nuclear crossroads :
At the nuclear crossroads : choices about nuclear
weapons and extension of the non-proliferation treaty
/ edited by John B. Rhinelander and Adam M. Scheinman.
-- Lanham : University Press of America : Lawyers
Alliance for World Security, c1994.
    p.   cm.
    Includes bibliographical references.
    ISBN 0-8191-9817-X (acid-free). -- ISBN
0-8191-9818-8 (pbk. : acid-free)
    1. Nuclear nonproliferation. 2. United States-
-Military policy. I. Rhinelander, John B.
II. Scheinman, Adam M.
JX1974.73.A8   1994
327.1'74--dc21 95-3155
CIP 2/95

Averting a Latin American nuclear arms race :
Averting a Latin American nuclear arms race : new
prospects and challenges for Argentine-Brazil nuclear
co-operation / edited by Paul L. Leventhal and Sharon
Tanzer. -- New York : St. Martin's Press ; Washington,
DC : Nuclear Control Institute, 1992.
    xiii, 257 p. : ill. ; 23 cm.
    Proceedings from the Conference on Latin-American
Nuclear Co-operation held in Montevideo, Uruguay,
October 11-13, 1989.
    Includes bibliographical references (p. 230-242) and
index.
    ISBN 0-312-07277-5
    1. Nuclear arms control--Argentina--Congresses.
2. Nuclear arms control--Brazil--Congresses. 3. Arms
race--Argentina--Congresses. 4. Arms race--Brazil-
-Congresses. 5. Nuclear nonproliferation--Congresses.
I. Leventhal, Paul. II. Tanzer, Sharon, 1941-   .
III. Conference on Latin-American Nuclear Co-operation
(1989 : Montevideo, Uruguay)
JX1974.7.A94   1992
327.1'74'098--dc20 91-31595

Avoiding nuclear anarchy :
Avoiding nuclear anarchy : containing the threat of
loose Russian nuclear weapons and fissile material /
Graham T. Allison ... [et al.]. -- Cambridge, Mass. :
MIT Press, c1996.
    viii, 295 p. : map ; 24 cm. -- (CSIA studies in
international security ; no. 12)
    Includes bibliographical references.
    ISBN 0-262-51088-X (pbk.)
    1. Nuclear terrorism--Prevention. 2. Nuclear
industry--Security measures--Former Soviet republics.
3. Nuclear nonproliferation--Government policy--United

Avoiding nuclear anarchy : (Cont'd)
Avoiding nuclear anarchy : containing the ... (Cont'd)
States.   I. Allison, Graham T.   II. Series.
HV6431.A96   1996
327.1'74--dc20   95-25376

Bailey, Kathleen C.
Bailey, Kathleen C.
   Doomsday weapons in the hands of many : the arms
control challenge of the '90s / Kathleen C. Bailey. --
Urbana : University of Illinois Press, c1991.
   158 p. : 24 cm.
   Includes bibliographical referencess and index.
   ISBN 0-252-01826-5 (cl). -- ISBN 0-252-06197-7 (pb)
   1. Nuclear arms control. 2. Chemical arms control.
3. Biological arms control. 4. Nuclear
nonproliferation.   I. Title.
JX1974.7.B34   1991
327.1'74--dc20   91-8081

Bailey, Kathleen C.
Bailey, Kathleen C.
   Strengthening nuclear nonproliferation / Kathleen C.
Bailey. -- Boulder : Westview Press, 1993.
   ix, 132 p. ; 24 cm.
   Includes bibliographical references and index.
   ISBN 0-8133-2006-2 (hardbound : acid-free paper). --
ISBN 0-8133-2007-0 (pbk. : acid-free paper)
   1. Nuclear nonproliferation.   I. Title.
JX1974.73.B35   1993
327.1'74--dc20   93-29169

Baker, John C., 1949-
Baker, John C., 1949-
   Non-proliferation incentives for Russia and Ukraine
/ John C. Baker. -- Oxford, [England] ; New York, NY :
Oxford University Press for the International
Institute for Strategic Studies, 1997.
   91 p. : ill. ; 24 cm. -- (Adelphi paper, ISSN
0567-932X ; 309)
   Includes bibliographical references (p. [85]-91).
   ISBN 0-19-829371-2
   1. Nuclear nonproliferation. 2. Arms transfers-
-Russia (Federation). 3. Arms transfers--Ukraine.
4. Defense industries--Russia (Federation).
5. Defense industries--Ukraine.   I. International
Institute for Strategic Studies.   II. Title.
III. Title: Nonproliferation incentives for Russia and
Ukraine.   IV. Series: Adelphi papers ; no. 309.
U162.A3   no.309
327.1'747'0947--dc21   98-101324

Barnaby, Frank.
Barnaby, Frank.
   How nuclear weapons spread : nuclear-weapon
proliferation in the 1990s / Frank Barnaby. -- London

Barnaby, Frank. (Cont'd)
Barnaby, Frank. (Cont'd)
; New York : Routledge, 1993.
xiii, 144 p. ; 24 cm. -- (Operational level of war)
Includes bibliographical references (p. 143-144) and index.
ISBN 0-415-07674-9
1. Nuclear nonproliferation.   I. Title.
II. Series.
JX1974.73.B37   1993
327.1'74--dc20   92-45847

Beck, Harald, 1960-
Beck, Harald, 1960-
Much ado about nothing? : the verification system of the South Pacific Nuclear Free Zone Treaty / by Harald Beck. -- Canberra : Australian National University, Research School of Pacific Studies, Peace Research Centre, c1991.
24 p. ; 30 cm. -- (Working paper, ISSN 0817-1831 ; no. 109)
"June 1991."
Includes bibliographical references.
ISBN 0-7315-1198-0
1. South Pacific Nuclear Free Zone Treaty (1985)
2. Nuclear-weapon-free zones--Pacific Area.
3. Nuclear nonproliferation.   I. Title.   II. Series: Working paper (Australian National University. Peace Research Centre) ; no. 109.
JX1974.74.P33B43   1991
341.7'34'0265--dc20   92-196738

Bee, Ronald J.
Bee, Ronald J.
Nuclear proliferation : the post-cold-war challenge / by Ronald J. Bee. -- New York, N.Y. : Foreign Policy Association, 1995.
72 p. : ill., map ; 20 cm. -- (Headline series, ISSN 0017-8780 ; no. 303)
Includes bibliographical references (p. 70-72).
ISBN 0-87124-160-9
1. Nuclear nonproliferation.   2. Nuclear weapons (International law)   I. Title.   II. Series.
JX1974.73.B44   1995
327.1'74--dc20   94-61086

Bernauer, Thomas.
Bernauer, Thomas.
Nuclear issues on the agenda of the Conference on Disarmament / Thomas Bernauer. -- New York : United Nations, 1991.
viii, 109 p. ; 30 cm.
At head of title: UNIDIR, United Nations Institute for Disarmament Research, Geneva.
"UNIDIR/91/68."
"United Nations publication sales no. GV.E.91.0.16"--T.p. verso.

Bernauer, Thomas. (Cont'd)
  Bernauer, Thomas. (Cont'd)
    Includes bibliographical references.
    ISBN 929045055X
    1. Nuclear disarmament. 2. Nuclear
  nonproliferation. 3. Nuclear arms control.
  I. United Nations Institute for Disarmament Research.
  II. Title.
    JX1974.7.B465   1991
    327.1'74--dc20   92-196063

Beyond 1995 :
  Beyond 1995 : the future of the NPT regime / edited by
  Joseph Pilat and Robert E. Pendley ; foreword by Hans
  M. Blix. -- New York : Plenum Press, c1990.
    xvii, 257 p. ; 24 cm. -- (Issues in international
  security)
    Includes bibliographical references and index.
    ISBN 0-306-43291-9
    1. Nuclear nonproliferation. 2. Security,
  International. I. Pilat, Joseph F. II. Pendley,
  Robert E. III. Series.
    JX1974.73.B49   1990
    327.1'74--dc20   89-29827

Bhatia, Shyam, 1950-
  Bhatia, Shyam, 1950-
    Nuclear rivals in the Middle East / Shyam Bhatia. --
  London ; New York : Routledge, 1988.
    119 p. ; ill. ; 22 cm.
    Bibliography: p. 113-114.
    Includes index.
    ISBN 0-415-00479-9
    1. Nuclear arms control--Middle East. 2. Arms race--Middle East. 3. Nuclear fuels--Middle East.
  4. Nuclear nonproliferation. I. Title.
    JX1974.74.M627B53   1988
    327.1'74'0956--dc19   87-33660

Bhola, P. L.
  Bhola, P. L.
    Pakistan's nuclear policy / P.L. Bhola. -- New Delhi
  : Sterling Publishers, c1993.
    viii, 222 p. ; 22 cm. -- (South Asia studies series)
    I-E-72555
    Includes bibliographical references (p. [210]-219)
  and index.
    ISBN 8120714822 : Rs200.00
    1. Nuclear energy--Government policy--Pakistan.
  2. Nuclear nonproliferation. I. Title. II. Series:
  South Asian studies series.
    QC773.3.P35B48   1993
    355.02'17'095491--dc20   93-908159

Bitzinger, Richard.
Bitzinger, Richard.
 The globalization of arms production : defense markets in transition / by Richard A. Bitzinger. -- Washington, D.C. : Defense Budget Project (777 North Capitol St., N.E., Suite 710, Washington, D.C. 20002), [1993]
  iii, 52 p. : ill. ; 28 cm.
  "December 1993."
  Includes bibliographical references.
  1. Defense industries. 2. Arms transfers.
  3. National security. 4. Nuclear nonproliferation.
  I. Center on Budget and Policy Priorities (Washington, D.C.). Defense Budget Project. II. Title.
  HD9743.A2B54 1993   96-170499

Blank, Stephen, 1950-
Blank, Stephen, 1950-
 Proliferation and nonproliferation in Ukraine : implications for European and U.S. security / Stephen J. Blank. -- Carlisle Barracks, PA : Strategic Studies Institute, U.S. Army War College, 1994.
  vi, 37 p. ; 23 cm.
  "July 1, 1994."
  Includes bibliographical references (p. 30-37).
  1. Nuclear arms control--Ukraine. 2. Ukraine--Military policy. 3. Ukraine--Strategic aspects.
  4. Nuclear nonproliferation. I. Army War College (U.S.). Strategic Studies Institute. II. Title.
  U264.5.U38B58 1994
  327.1'74'094771--dc20   94-201112

Blix, Hans.
Blix, Hans.
 Can we stop the spread of nuclear weapons? / Hans Blix. -- Vienna, Austria : International Atomic Energy Agency, [1990]
  13 p. ; 21 cm.
  1. Nuclear nonproliferation. I. Title.
  JX1974.73.B545 1990
  327.1'74--dc20   90-225723

Blocking the spread of nuclear weapons :
 Blocking the spread of nuclear weapons : American and European perspectives. -- New York : Council on Foreign Relations, c1986.
  x, 153 p. ; 23 cm.
  "Published in cooperation with the Centre for European Policy Studies, CEPS."
  ISBN 0-87609-012-9 (pbk.) : $6.95
  1. Nuclear nonproliferation. I. Council on Foreign Relations. II. Centre for European Policy Studies (Louvain-la-Neuve, Belgium)

Blocking the spread of nuclear weapons : (Cont'd)
Blocking the spread of nuclear weapons : ... (Cont'd)
JX1974.73.B55   1986
327.1'74--dc19   86-4196

Booker, Malcolm.
   Booker, Malcolm.
   Nuclear war : present and future dangers / Malcolm
Booker. -- Sydney : Left Book Club Co-operative, 1993.
   vi, 58 p. ; 21 cm.
   Includes bibliographical references (p. 58).
   ISBN 1-875285-13-X
   1. Nuclear arms control. 2. Nuclear
nonproliferation.   I. Title.
JX1974.7.B636   1993
327.1'74--dc20   93-203659

Bridging the nonproliferation divide :
   Bridging the nonproliferation divide : the United States
and India / edited by Francine R. Frankel. -- Lanham,
Md. : University Press of America ; [Philadelphia,
Pa.] : Center for the Advanced Study of India, c1995.
   xiv, 363 p. : ill. (some col.) ; 24 cm.
   Includes bibliographical references.
   ISBN 0-8191-9943-5 (cloth : alk. paper). -- ISBN
0-8191-9944-3 (pkb. : alk. paper)
   1. Nuclear nonproliferation. 2. Nuclear arms
control--India. 3. Nuclear arms control--United
States.   I. Frankel, Francine R.
JX1974.73.B75   1995
327.1'74--dc20   95-14734

Bundy, McGeorge.
   Bundy, McGeorge.
   Reducing nuclear danger : the road away from the
brink / McGeorge Bundy, William J. Crowe, Jr., Sidney
D. Drell. -- New York : Council on Foreign Relations
Press, c1993.
   ix, 107 p. ; 23 cm.
   "Council on Foreign Relations books"--T.p. verso.
   Includes bibliographical references and index.
   ISBN 0-87609-149-4 : $14.95
   1. United States--Military policy. 2. Nuclear
weapons--United States. 3. Nuclear nonproliferation.
4. Nuclear arms control.   I. Crowe, William J., 1925-
.   II. Drell, Sidney D. (Sidney David), 1926-   .
III. Title.
UA23.B7864   1993
355'.0335'73--dc20   93-35831
                                                                r94

Bunn, George, 1925-
   Bunn, George, 1925-
   Extending the non-proliferation treaty : legal
questions faced by the parties in 1995 / by George
Bunn. -- Washington, D.C. : American Society of

Bunn, George, 1925- (Cont'd)
Bunn, George, 1925-   (Cont'd)
International Law, c1994.
   viii, 56 p. ; 28 cm. -- (Issue papers on world
conferences ; no. 2)
   Includes bibliographical references.
   1. Nuclear nonproliferation. 2. Nuclear weapons
(International law) 3. Treaty on the Non
-proliferation of Nuclear Weapons (1968)   I. Title.
II. Series.
JX1974.73.B86   1994
341.7'34--dc20   95-202646

Bunn, T. Davis, 1952-
Bunn, T. Davis, 1952-
   Riders of the pale horse / T. Davis Bunn. --
Minneapolis, Minn. : Bethany House, c1994.
   348 p. : maps ; 22 cm.
   ISBN 1-55661-346-6 : $9.99
   1. Islamic fundamentalism--Middle East--Fiction.
2. Nuclear nonproliferation--Fiction. 3. Americans-
-Middle East--Fiction.   I. Title.
PS3552.U4718R53   1994
813'.54--dc20   93-45734
                                                 r96

Canberra Commission on the Elimination of Nuclear Weapons.
Canberra Commission on the Elimination of Nuclear
Weapons.
   Executive summary : report of the Canberra
Commission on the Elimination of Nuclear Weapons. --
[Canberra] : Dept. of Foreign Affairs and Trade,
c1996.
   24 p. : ill. ; 25 cm.
   "August 1996.
   ISBN 0-642-25091-X
   1. Nuclear disarmament--Australia. 2. Nuclear
nonproliferation.   I. Australia. Dept. of Foreign
Affairs and Trade. II. Title.
JZ5665.C36   1996
327.1'747'0994--dc21   97-155350

Central European countries and non-proliferation regimes /
Central European countries and non-proliferation regimes
/ Harald Müller, Janusz Prystrom, (eds). -- Warsaw :
Polish Foundation of International Affairs ;
[Frankfurt am Main] : Peace Research Institute
Frankfurt, c1996.
   156 p. : ill. ; 24 cm.
   Includes bibliographical references.
   ISBN 8390551454
   1. Nuclear nonproliferation. 2. Nuclear industry-
-Law and legislation--Europe, Eastern. 3. Export
controls--Europe, Eastern.   I. Müller, Harald, 1949
May 13- II. Prystrom, Janusz. III. Polish Foundation
of International Affairs. IV. Hessische Stiftung
Friedens- und Konfliktforschung.

Central European countries and ... (Cont'd)
Central European countries and ... (Cont'd)
JZ5675.C443   1996
327.1'747'0943   97-190163

Chauvistré, Eric.
Chauvistré, Eric.
   Germany and proliferation : the nuclear export
policy / Eric Chauvistré. -- Berlin : Berghof-Stiftung
für Konfliktforschung, 1991.
   48 leaves ; 30 cm. -- (Arbeitspapiere der Berghof
-Stiftung für Konfliktforschung, ISSN 0936-6857 ; Nr.
43)
   Abstract in English and German.
   Includes bibliographical references.
   1. Nuclear industry--Government policy--Germany
(West) 2. Nuclear nonproliferation.   I. Title.
II. Series.
HD9698.G42C47   1991   92-135262

Chellaney, Brahma.
Chellaney, Brahma.
   Nuclear proliferation : the U.S.-Indian conflict /
Brahma Chellaney. -- New Delhi : Orient Longman, 1993.
   xi, 474 p. ; 23 cm.
   I-E-71414
   Includes bibliographical references (p. [412]-465)
and index.
   ISBN 0-86311-327-3 : Rs300.00
   1. Nuclear nonproliferation.   2. United States-
-Foreign relations--India.   3. India--Foreign
relations--United States.   I. Title.
JX1974.73.C48   1993
327.73054--dc20   93-902444

Cheney, Glenn Alan.
Cheney, Glenn Alan.
   Nuclear proliferation : the problems and
possibilities / Glenn Alan Cheney. -- New York :
Franklin Watts, 1999.
         p.   cm. -- (Impact book)
   Includes bibliographical references and index.
   Summary: Discusses current policies governing the
spread of nuclear weapons along with potential
problems and possible outcomes.
   ISBN 0-531-11431-7
   1. Nuclear nonproliferation--Juvenile literature.
2. Nuclear weapons--Juvenile literature.
3. Radioactive substances--Juvenile literature.
[1. Nuclear nonproliferation.   2. Nuclear weapons.
3. Disarmament.   4. Radioactive substances.]
I. Title.
JZ5675.C45   1999
327.1'747--dc21   98-4526
                                                CIP 3/99
                                                     AC

Chow, Brian G.
 Chow, Brian G.
  The proposed fissile-material production cutoff :
  next steps / Brian G. Chow, Richard H. Speier, Gregory
  S. Jones. -- Santa Monica, CA : Rand, 1995.
   xv, 51 p. : ill. ; 23 cm.
   "National Defense Research Institute."
   "Prepared for the Office of the Secretary of
  Defense."
   Includes bibliographical references.
   ISBN 0-8330-1654-7
   1. Nuclear nonproliferation. 2. Nuclear
  disarmament--United States. I. Speier, Richard.
  II. Jones, Gregory S. III. United States. Dept. of
  Defense. Office of the Secretary of Defense.
  IV. National Defense Research Institute (U.S.)
  V. Title.
  JX1974.73.C49   1995
  327.1'74--dc20   95-12856

Cole, Paul M.
 Cole, Paul M.
  Sweden without the bomb : the conduct of a nuclear
  -capable nation without nuclear weapons / Paul M.
  Cole. -- Santa Monica, CA : Rand, 1994.
   xxi, 273 p. ; 28 cm.
   "Prepared for the Office of Research and
  Development, Central Intelligence Agency."
   "National Security Research Division."
   Includes bibliographical references (p. 257-273).
   ISBN 0-8330-1583-4
   1. National security--Sweden. 2. Nuclear weapons-
  -Sweden. 3. Nuclear nonproliferation. 4. Sweden-
  -Foreign relations--United States. 5. Sweden-
  -Politics and government. 6. United States--Foreign
  relations--Sweden. I. United States. Central
  Intelligence Agency. Office of Research and
  Development. II. Rand Corporation. National Security
  Research Division. III. Title.
  UA790.C65   1994
  355.02'17'09485--dc20   95-120662
                                                     r962

Comprehensive disclosure of fissionable materials :
 Comprehensive disclosure of fissionable materials : a
  suggested initiative / Carnegie Commission on
  Preventing Deadly Conflict. -- Washington, D.C. :
  Carnegie Corp. of New York : Additional copies of this
  report may be obtained free of charge from the
  commission's headquarters, Carnegie Commission on
  Preventing Deadly Conflict, 1995.
   iii, 11 p. ; 26 cm. -- (Discussion paper / Carnegie
  Commission on Preventing Deadly Conflict)
   "June 1995."
   1. Nuclear nonproliferation. 2. Nuclear fuels.

Comprehensive disclosure of fissionable ... (Cont'd)
 Comprehensive disclosure of fissionable ... (Cont'd)
 3. Disclosure of information. I. Carnegie Commission
 on Preventing Deadly Conflict. II. Series: Discussion
 paper (Carnegie Commission on Preventing Deadly
 Conflict)
 KZ5684.C65 1995   97-139016

Constraining proliferation :
 Constraining proliferation : the contribution of
 verification synergies / by Patricia Bliss McFate ...
 [et al.] ; prepared for the Non-Proliferation, Arms
 Control, and Disarmament Division, Department of
 External Affairs, Ottawa Canada. -- Ottawa : External
 Affairs and International Trade Canada, c1993.
   v, 70 p. : ill. ; 28 cm. -- (Arms control
 verification studies, ISSN 0828-3664 ; no. 5)
   C93-99452-3
   Includes bibliographical references.
   Abstract in French.
   DSS Cat. no. E54-5/5E
   ISBN 0-662-20347-X
   1. Nuclear nonproliferation. 2. Nuclear arms
 control--Verification. I. McFate, Patricia Bliss.
 II. Canada. Non-Proliferation, Arms Control and
 Disarmament Division. III. Series.
 JX1974.73.C658 1993   94-182189

Control but verify :
 Control but verify : verification and the new non
 -proliferation agenda / edited by David Mutimer. --
 Toronto, Canada : Centre for International and
 Strategic Studies, York University, 1994.
   xii, 227 p. ; 23 cm.
   C94-931130-8
   Papers presented at, or commissioned for, two
 workshops held in Ottawa, Ont., Dec. 21-22, 1992, and
 Nov. 25-26, 1993.
   Includes bibliographical references.
   ISBN 0-920231-17-9
   1. Arms control--Verification. 2. Nuclear
 nonproliferation. I. Mutimer, David, 1964-   .
 II. York Centre for International and Strategic
 Studies.
 JX1974.C692 1994
 327.1'74--dc20   95-152507

Controlling the atom in the 21st century /
 Controlling the atom in the 21st century / edited by
 David P. O'Very, Christopher E. Paine, and Dan W.
 Reicher. -- Boulder : Westview Press, 1994.
   xiv, 397 p. ; 23 cm.
   Includes bibliographical references and index.
   ISBN 0-8133-8816-3 (alk. paper)
   1. Nuclear energy--Law and legislation. 2. Nuclear
 power plants--Licenses. 3. Nuclear nonproliferation.
 I. O'Very, David P. II. Paine, Christopher E.

Controlling the atom in the 21st century / (Cont'd)
Controlling the atom in the 21st century / ... (Cont'd)
III. Reicher, Dan W.
K3990.6.C66 1994
333.792'4--dc20   93-39042
                                                   r94

CSIS U.S.-Euratom Senior Policy Panel.
   CSIS U.S.-Euratom Senior Policy Panel.
      Negotiating a U.S.-Euratom successor agreement : finding common ground in nuclear cooperation : a consensus report of the CSIS U.S.-Euratom Senior Policy Panel / project chairman, James R. Schlesinger ; CSIS U.S.-Euratom Senior Policy Panel, David M. Abshire ... [et al.] ; project director, Dan Gouré ; project coordinators, Alexander T. Lennon, David C. Earnest. -- Washington, D.C. : Center for Strategic and International Studies, c1994.
      v, 33 p. ; 28 cm. -- (Panel report, ISSN 0899-0352)
      Includes bibliographical references (p. 27-29).
      ISBN 0-89206-258-4
      1. Nuclear industry--Government policy--United States. 2. Nuclear industry--Government policy--Europe. 3. Nuclear nonproliferation--International cooperation. 4. Euratom. I. Title. II. Series: CSIS panel reports.
JX1974.73.C77 1994
327.1'74--dc20   94-39871

Culture and security :
   Culture and security : multilateralism, arms control, and security building / edited by Keith R. Krause. -- Portland, Ore : Frank Cass, 1998.
      p.   cm.
      "This group of studies first appeared as a special issue of Contemporary security policy, Vol. 19, No. 1, April 1998."
      Includes bibliographical references and index.
      ISBN 0-7146-4885-X (cloth). -- ISBN 0-7146-4437-4 (pbk.)
      1. International relations. 2. Security, International. 3. Nuclear nonproliferation. 4. International relations and culture. I. Krause, Keith. II. Contemporary security policy.
JZ1242.C85 1998
327.1'747--dc21   98-25998
                                                 CIP 9/98

Davydov, Valery.
   Davydov, Valery.
      An annotated bibliography of Soviet and CIS studies on nuclear nonproliferation. -- Monterey, CA : Center for Russian and Eurasian Studies, Monterey Institute of International Studies, c1992.
      16 p. ; 28 cm. -- (Occasional paper ; no. 1)
      ISBN 0-9633859-0-9
      1. Nuclear nonproliferation--Bibliography.

Davydov, Valery. (Cont'd)
Davydov, Valery. (Cont'd)
2. Nuclear arms control--Soviet Union--Bibliography.
I. Title. II. Series: Occasional paper (Center for
Russian and Eurasian Studies (Monterey Institute of
International Studies)); no. 1.
Z6464.D6D29 1992
016.3417'34--dc20   92-242933

Dembinski, Matthias.
Dembinski, Matthias.
NATO and nonproliferation : a critical appraisal /
Matthias Dembinski, Alexander Kelle, Harald Müller. --
Frankfurt am Main : Peace Research Institute
Frankfurt, [1994]
iv, 51 p. ; 30 cm. -- (PRIF reports ; no. 33)
"April 1994."
ISBN 3-928965-39-5
1. Nuclear nonproliferation. 2. North Atlantic
Treaty Organization. I. Kelle, Alexander.
II. Müller, Harald, 1949 May 13- III. Title.
IV. Series.
JX1974.7.D44188 1994
327.1'747'094--dc21   95-177841

Dhanjal, Gursharan S.
Dhanjal, Gursharan S.
Tarapur, the politics of nuclear age / Gursharan S.
Dhanjal. -- Delhi : Rajdhani Book Service, 1992.
vi, 154 p. ; 22 cm.
I-E-70183
Summary: On the Indo-US nuclear power diplomacy with
special reference to the Nuclear Non-Proliferation
Treaty, 1968.
Includes bibliographical references (p. [143]-152)
and index.
Rs210.00
1. Nuclear nonproliferation. 2. United States-
-Foreign relations--India. 3. India--Foreign
relations--United States. I. Title.
JX1974.73.D49 1992
327.54073--dc20   92-909917

Disarmament :
Disarmament : assuring the success of the Non
-Proliferation Treaty extension conference : excerpts
from the panel discussions organized by the NGO
Committee on Disarmament, Inc., at the Conference held
at the United Nations in New York, 20-21 April 1994. -
- New York : United Nations, 1994.
121 p. ; 28 cm.
"United Nations publication sales no. E.94.IX.9."
ISBN 9211422094
1. Nuclear nonproliferation--Congresses. 2. Nuclear
disarmament--Congresses. 3. Security, International-
-Congresses. I. NGO Committee on Disarmament.

Disarmament : (Cont'd)
Disarmament : assuring the success of the ... (Cont'd)
JX1974.73.D57   1994
327.1'74--dc20   95-130442

Disarmament :
Disarmament : 1996--disarmament at a critical juncture : panel discussions organized by the NGO Committee on Disarmament. -- New York : United Nations, 1997.
141 p. ; 28 cm.
"Edited transcript of the presentations made by Joseph Rotblat, winner of the 1995 Nobel Peace Prize, and by Minoru Ohmuta and Hidehiko Yoko-o, speaking for the Mayors of Hiroshima and Nagasaki, at the United Nations on 25 April 1996; Excerpts from the panel discussion on South Asia at a Critical Turning Point, held on 24 September 1996; Extensive excerpts from the panel discussions held 22-24 October 1996 during Disarmament Week at the United Nations."
"United Nations publication sales no. E.97.IX.2."
ISBN 9211422205
1. Nuclear nonproliferation--Congresses. 2. Nuclear disarmament--Congresses. 3. Nuclear arms control--South Asia--Congresses. I. Rotblat, Joseph, 1908- . II. Ohmuta, Minoru. III. Yoko-o, Hidehiko. IV. NGO Committee on Disarmament.
JZ5675.D57   1997
327.1'747--dc21   97-213197

Disarmament and arms control :
Disarmament and arms control : a booklet prepared to mark Disarmament Week, 24-30 October 1985. -- Wellington, N.Z. : Ministry of Foreign Affairs, [1985]
53 p. : ill. ; 30 cm. -- (Information bulletin, ISSN 0111-8315 ; no. 13)
Cover title.
"October 1985."
1. Nuclear disarmament--New Zealand. 2. Nuclear disarmament. 3. Nuclear nonproliferation. I. New Zealand. Ministry of Foreign Affairs. II. Series: Information bulletin (New Zealand. Ministry of Foreign Affairs) ; no. 13.
JX1974.7.D56   1985
327.1'74'09931--dc19   86-143171

Dokos, Thanos P., 1965-
Dokos, Thanos P., 1965-
Non-proliferation and Greek foreign policy / Thanos Dokos. -- Athens : Hellenic Foundation for European and Foreign Policy, 1994.
10 p. ; 30 cm. -- (Defense and foreign policy studies = Meletes amyntikēs kai exōterikēs politikēs ; 14)
Cover title.
1. Nuclear nonproliferation. 2. Greece--Foreign relations--1974-   I. Title. II. Series: Defense and foreign policy studies ; 14.

Dokos, Thanos P., 1965- (Cont'd)
Dokos, Thanos P., 1965-   (Cont'd)
JX1974.73.D654 1994   96-135667
                                                    r97
Donnelly, Warren H.
  Donnelly, Warren H.
    The International Atomic Energy Agency : application
  of safeguards in the United States : an analysis of
  the agreement and an assessment of the negotiation /
  prepared for the Committee on Foreign Relations,
  United States Senate ; by the Environment and Natural
  Resources Policy Division, Library of Congress. --
  Washington : U.S. G.P.O., 1979.
    vi, 91 p. ; 24 cm.
    At head of title: 96th Congress, 1st session.
  Committee print.
    Authors: Warren H. Donnelly and Allan M. Labowitz.
    "May 1977."
    Includes bibliographical references.
    Item 1039
    Supt. of Docs. no.: Y 4.F 76/2:In 8/51
    1. Nuclear nonproliferation. 2. Nuclear facilities-
  -Law and legislation--United States. 3. International
  Atomic Energy Agency. I. Labowitz, Allan M.
  II. United States. Congress. Senate. Committee on
  Foreign Relations. III. Library of Congress.
  Environment and Natural Resources Policy Division.
  IV. Title.
  K3990.4.D66 1979
  327.1'74--dc19
  DGPO/DLC
     for Library of Congress   84-603507

Egypt and the peaceful uses of nuclear energy /
  Egypt and the peaceful uses of nuclear energy / Arab
  Republic of Egypt, Ministry of Foreign Affairs. --
  Cairo : The Ministry, 1984.
    319 p. ; 24 cm.
    Contents: Egyptian decrees and legislation
  concerning the regulation of the peaceful uses of
  nuclear energy -- Peaceful nuclear cooperation
  agreements in the fields of research, the exchange of
  technical equipment, information and nuclear safety --
  Multilateral agreements in the field of peaceful uses
  of nuclear energy -- Convention on Arab cooperation in
  the peaceful uses of nuclear energy -- The Egyptian
  Nuclear Power Program for the generation of
  electricity -- Bilateral agreements on cooperation in
  the peaceful uses of atomic energy -- Agreement
  between the Arab Republic of Egypt and the
  International Atomic Energy Agency for the application
  of safeguards in connection with the Treaty on
  Profileration of Nuclear Weapons.
    ISBN 9770208442
    1. Nuclear energy--Law and legislation--Egypt.
  2. Nuclear nonproliferation. I. Egypt. Wizarat al

Egypt and the peaceful uses of nuclear energy / (Cont'd)
Egypt and the peaceful uses of nuclear ... (Cont'd)
-Khārijīyah.
LAW <EGYPT 4 Nuclear 1984>   85-215405

Egypt and the Treaty on the Non-Proliferation of Nuclear Weapons /
Egypt and the Treaty on the Non-Proliferation of Nuclear Weapons / Arab Republic of Egypt, Ministry of Foreign Affairs. -- Cairo : State Information Service, 1981.
   110 p. ; 24 cm.
   Cover title: Egypt and treaty.
   Includes bibliographical references (p. 109-110).
   1. Treaty on the Non-Proliferation of Nuclear Weapons (1968) 2. Nuclear nonproliferation.
   3. Nuclear arms control--Egypt. I. Egypt. Wizārat al-Khārijīyah. II. Title: Egypt and treaty.
   JX1974.73.E35 1981   86-969107

Environmental monitoring for nuclear safeguards.
Environmental monitoring for nuclear safeguards. -- Washington, DC : Office of Technology Assessment, Congress of the U.S. : For sale by the U.S. G.P.O., Supt. of Docs., [1995]
   vii, 44 p. : ill. ; 28 cm.
   Shipping list no.: 96-0037-P.
   "September 1995"--P. [4] of cover.
   Includes bibliographical references.
   "OTA-BP-ISS-168"--P. [4] of cover.
   Supt. of Docs. no.: Y 3.T 22/2:2 M 74/2
   GPO: S/N 052-003-01442-4
   ISBN 0-16-048323-9 : $4.00
   1. International Atomic Energy Agency. 2. Nuclear arms control--Verification. 3. Nuclear nonproliferation. 4. Environmental monitoring.
   5. Nuclear weapons plants--Inspection. I. United States. Congress. Office of Technology Assessment.
   UA12.5.E58 1995
   327.1'747--dc21   95-228362

A European non-proliferation policy :
A European non-proliferation policy : prospects and problems / edited by Harald Müller. -- Oxford [Oxfordshire] : Clarendon Press ; New York : Oxford University Press, 1987.
   xxii, 416 p. ; 24 cm.
   Includes bibliographies and index.
   ISBN 0-19-829702-5 : $60.00 (U.S. : est.)
   1. Nuclear nonproliferation. 2. Nuclear arms control--Europe. 3. Europe--Military policy.
   I. Müller, Harald, 1949 May 13-
   JX1974.73.E87 1987
   327.1'74'094--dc19   87-12342
                                                   r90

European non-proliferation policy, 1988-1992 /
European non-proliferation policy, 1988-1992 / Harald Müller, ed. -- Brussels : European Interuniversity Press, 1993?
   259 p. ; 24 cm. -- (Collection "La cité européenne" ; no 4)
Includes bibliographical references.
ISBN 9052013055
1. Nuclear nonproliferation. 2. Nuclear weapons (International law) 3. Nuclear arms control--Europe. 4. Nuclear energy--Government policy--Europe. I. Müller, Harald, 1949 May 13- II. Series.
JX1974.73E87 1993
327.1'74--dc20   95-210310

Export controls and nonproliferation policy.
Export controls and nonproliferation policy. -- Washington, DC : Office of Technology Assessment, U.S. Congress : For sale by the U.S. G.P.O., Supt. of Docs., [1994]
   ix, 82 p. : ill. ; 28 cm.
Shipping list no.: 94-0187-P.
"May 1994."
Includes bibliographical references.
Includes index.
"OTA-ISS-596"--P. [4] of cover.
S/N 052-003-01371-1 (GPO)
Supt. of Docs. no.: Y 3.T 22/2:EX 7/2
GPO: 1070-M
ISBN 0-16-045014-4 : $5.50
1. Nuclear nonproliferation. 2. Export controls--United States. 3. Technology transfer--Law and legislation--United States. I. United States. Congress. Office of Technology Assessment.
JX1974.73.E98 1994
382'.4562345119'0973--dc20
DGPO/DLC
for Library of Congress   94-190230

Feldman, Shai, 1950-
Feldman, Shai, 1950-
   Nuclear weapons and arms control in the Middle East / by Shai Feldman. -- Cambridge, Mass. : MIT Press, c1997.
   xvii, 336 p. ; 24 cm. -- (CSIA studies in international security)
Includes bibliographical references and index.
ISBN 0-262-06189-9 (alk. paper). -- ISBN 0-262-56108-5 (pbk. : alk. paper)
   1. Nuclear arms control--Middle East. 2. Nuclear nonproliferation. 3. Security, International. I. Title. II. Series.
JX1974.74.M627F45 1997
327.1'747'0956--dc21   96-44081

Fighting proliferation :
Fighting proliferation : new concerns for the nineties / edited by Henry Sokolski. -- Maxwell Air Force Base, Ala. : Air University Press ; Washington, D.C. : For sale by the Supt. of Docs., US G.P.O., [1996]
    xix, 377 p. : ill. ; 24 cm.
    "September 1996."
    Includes bibliographical references and index.
    1. Nuclear nonproliferation.   I. Sokolski, Henry.
    JX1974.73.F54   1996
    327.1'74--dc20   96-24696

Fischer, David, 1920-
Fischer, David, 1920-
    Safeguarding the atom : a critical appraisal / David Fischer and Paul Szasz ; edited by Jozef Goldblat. -- London ; Philadelphia : Taylor & Francis, 1985.
    xx, 243 p. : ill. ; 24 cm.
    "Stockholm International Peace Research Institute."
    Includes bibliographical references and index.
    ISBN 0-85066-306-7
    1. Nuclear nonproliferation.   I. Szasz, Paul C. II. Goldblat, Jozef.   III. Stockholm International Peace Research Institute.   IV. Title.
    JX1974.73.F58   1985
    327.1'74--dc19   85-181278
                                                        r96

Fischer, David, 1920-
Fischer, David, 1920-
    Stopping the spread of nuclear weapons : the past and the prospects / David Fischer. -- London ; New York : Routledge, 1992.
    336 p. ; 22 cm.
    Includes bibliographical references (p. [250]-313) and index.
    ISBN 0-415-00481-0
    1. Nuclear nonproliferation.   I. Title.
    JX1974.73.F59   1992
    327.1'74--dc20   91-10061

Fischer, David, 1920-
Fischer, David, 1920-
    Towards 1995 : the prospects for ending the proliferation of nuclear weapons / David Fischer. -- [Geneva] : UNIDIR ; Aldershot ; Brookfield, Vt., USA : Dartmouth, c1993.
    vii, 292 p. : map ; 26 cm.
    Includes bibliographical references.
    ISBN 1-85521-322-2 : $59.95 (est.)
    1. Nuclear nonproliferation.   I. Title.
    JX1974.F469   1993
    327.1'74--dc20   92-33233
                                                        r93

Fischer, David, 1920-
Fischer, David, 1920-
A treaty in trouble : Europe and the NPT after the Fourth Review Conference / David Fischer, Harald Müller. -- Frankfurt am Main : Peace Research Institute Frankfurt, [1991]
ii, 44 p. ; 30 cm. -- (PRIF reports ; no. 17)
Includes bibliographical references.
ISBN 3-926197-84-6 : DM18.00 ($12.50 U.S.)
1. Nuclear nonproliferation--Congresses. 2. Nuclear energy--Government policy--European Economic Community countries--Congresses. I. Müller, Harald, 1949 May 13- II. Hessische Stiftung Friedens- und Konfliktforschung. III. Title. IV. Series.
JX1974.73.F592   1991
327.1'74--dc20   91-141407
                                                             r93

Fischer, David, 1920-
Fischer, David, 1920-
United divided : the European at the NPT Extension Conference / David Fischer [and] Harald Müller. -- Frankfurt/Main : Peace Research Institute Frankfurt, [1995]
i leaf, 49 p. ; 30 cm. -- (PRIF reports ; no. 40)
Includes bibliographical references.
"November 1995."
ISBN 3-928965-63-8
1. Nuclear nonproliferation. 2. Nuclear arms control--European Union countries. I. Müller, Harald, 1949 May 13- II. Non-Proliferation Treaty Extension Conference (1994 : United Nations) III. Hessische Stiftung Friedens- und Konfliktforschung. IV. Title. V. Series.
JZ5675.F57   1995
341.7'34--dc21   96-202430

Freeman, J. P. G. (John Patrick George)
Freeman, J. P. G. (John Patrick George)
Britain's nuclear arms control policy in the context of Anglo-American relations, 1957-68 / J.P.G. Freeman ; foreword by Laurence Martin. -- New York : St. Martin's Press, 1986.
xvi, 317 p. ; 23 cm.
Bibliography: p. 301-310.
Includes index.
ISBN 0-312-09959-2 : $29.95
1. Nuclear arms control--Great Britain--History. 2. Nuclear nonproliferation--History. 3. Great Britain--Foreign relations--United States. 4. United States--Foreign relations--Great Britain. I. Title.
JX1974.7.F733   1986
327.1'74'0941--dc19   85-24998

Fujita, Edmundo.
Fujita, Edmundo.
The prevention of geographical proliferation of
nuclear weapons : nuclear-weapon-free zones and zones
of peace in the Southern Hemisphere / Edmundo Fujita.
-- New York : United Nations, 1989.
iii, 57 p. : maps ; 30 cm. -- (Research paper, ISSN
1014-4013 ; no. 4)
At head of title: UNIDIR, United Nations Institute
for Disarmament Research.
"UNIDIR/88/24."
"United Nations publication sales no. GV. E.89.0.8"-
-T.p. verso.
Includes bibliographical references (p. 41-42).
ISBN 9290450347
1. Nuclear-weapon-free zones--Southern Hemisphere.
2. Nuclear nonproliferation. 3. Zones of peace-
-Southern Hemisphere. I. United Nations Institute
for Disarmament Research. II. Title. III. Series:
Research paper (United Nations Institute for
Disarmament Research) ; no. 4.
JX1974.735.F85   1989   90-225371
                                                      r96

The future of the international nuclear non-proliferation regime /
The future of the international nuclear non
-proliferation regime / edited by Marianne van
Leeuwen. -- Dordrecht ; Boston : M. Nijhoff
Publishers, c1995.
vi, 326 p. ; 24 cm. -- (Nijhoff law specials ; 10)
At head of title: Netherlands Institute of
International Relations "Clingendael."
Includes bibliographical references and index.
ISBN 0-7923-3433-7 (pbk. : acid-free paper)
1. Nuclear nonproliferation. 2. Nuclear arms
control. I. Leeuwen, M. van (Marianne)
II. Nederlands Instituut voor Internationale
Betrekkingen "Clingendael." III. Series.
JX1974.73.F877   1995
327.1'74--dc20   95-11565
                                                     r97

The future of the Non-proliferation Treaty /
The future of the Non-proliferation Treaty / edited by
John Simpson and Darryl Howlett. -- New York, N.Y. :
St. Martin's Press, 1995.
xiv, 226 p. ; 23 cm.
Papers presented at a July 1993 seminar organized by
PPNN in Southampton, England.
Includes bibliographical references (p. 207-211) and
index.
ISBN 0-312-12279-9. -- ISBN 0-333-61857-2
1. Treaty on the Non-proliferation of Nuclear
Weapons (1968) 2. Nuclear nonproliferation.

The future of the Non-proliferation Treaty / (Cont'd)
The future of the Non-proliferation Treaty ... (Cont'd)
I. Simpson, John, 1943-   . II. Howlett, Darryl A.,
1954-   .
JX1974.73.F88   1995
341.7'34--dc20   94-18289
                                                            r95

Gardner, Gary T., 1958-
Gardner, Gary T., 1958-
    Nuclear nonproliferation : a primer / Gary T.
Gardner. -- Boulder : L. Rienner Publishers, 1994.
    xiii, 141 p. : ill. ; 24 cm.
    "Published in association with the Program for
Nonproliferation Studies, Monterey Institute of
International Studies"--P. facing t.p.
    Includes bibliographical references (p. 127-130) and
index.
    ISBN 1-55587-478-9 (alk. paper). -- ISBN
1-55587-489-4 (pbk. : alk. paper)
    1. Nuclear nonproliferation.   I. Title.
JX1974.73.G374   1994
327.1'74--dc20   93-23571

Global problems of mankind and the state :
    Global problems of mankind and the state : a theme of
the 13th World Congress of the International Political
Science Association (Paris, July 1985). -- Moscow :
"Social Sciences Today" Editorial Board, USSR Academy
of Sciences, 1985.
    216 p. ; 22 cm. -- (Problems of the contemporary
world ; no. 116)
    Includes bibliographies.
    0.80rub (pbk.)
    1. State, The--Congresses. 2. World politics--1945-
--Congresses. 3. International relations--Congresses.
4. Nuclear nonproliferation--Congresses. 5. Peace-
-Congresses.   I. International Political Science
Association. World Congress (13th : 1985 : Paris,
France) II. Series.
JC348.G56   1985
320.1--dc19   85-229723

Goldblat, Jozef.
Goldblat, Jozef.
    The non-proliferation treaty : how to remove the
residual threats / Jozef Goldblat. -- New York :
United Nations, 1992.
    v, 36 p. ; 30 cm. -- (Research paper, ISSN 1014-4013
; no. 13)
    At head of title: UNIDIR, United Nations Institute
for Disarmament Research, Geneva.
    "United Nations publication sales no. GV.E.92.0.25."
    "UNIDIR/92/64."
    ISBN 929045069X
    1. Nuclear nonproliferation. 2. Nuclear arms
control--Verification.   I. United Nations Institute

Goldblat, Jozef. (Cont'd)
Goldblat, Jozef. (Cont'd)
for Disarmament Research.   II. Title.   III. Series:
Research paper (United Nations Institute for
Disarmament Research) ; no. 13.
JX1974.73.G6494 1992   94-120325

Goldblat, Jozef.
Goldblat, Jozef.
Nuclear non-proliferation : a guide to the debate /
Jozef Goldblat. -- London ; Philadelphia : Taylor &
Francis, 1985.
viii, 95 p. ; 23 cm.
"SIPRI, Stockholm International Peace Research
Institute."
"Prepared for ... third Review Conference of the
parties to the Treaty on the Non-Proliferation of
Nuclear Weapons to be held in September 1985"--P. v.
"Summarizes the findings reached in a study of the
non-proliferation policies of 15 countries ... in Non
-proliferation : the why and the wherefore"--P. v.
ISBN 0-85066-310-5 (pbk.)
1. Nuclear nonproliferation.   I. Stockholm
International Peace Research Institute.   II. Non
-Proliferation Treaty Review Conference (3rd : 1985 :
Geneva, Switzerland)   III. Non-proliferation.
IV. Title.
JX1974.73.G65 1985
327.1'74--dc19   85-12589
                                                  r86

Goldblat, Jozef.
Goldblat, Jozef.
Twenty years of the Non-proliferation Treaty :
implementation and prospects / by Jozef Goldblat. --
Oslo : PRIO, International Peace Research Institute,
c1990.
162 p. ; 21 cm.
Includes bibliographical references (p. 151-158).
ISBN (invalid) 82728810041470
1. Nuclear nonproliferation.   I. International
Peace Research Institute.   II. Title.
JX1974.73.G66 1990
327.1'74--dc20   90-186087

Green, Alex Edward Samuel, 1919-
Green, Alex Edward Samuel, 1919-
Defense conversion : a critical East-West experiment
/ edited by Alex E.S. Green ; with former Soviet Union
perspectives by Victor V. Chernyy. -- Hampton, Va.,
USA ; A. Deepak Pub., 1995.
xxiii, 337 p. : ill. ; 24 cm. -- (Studies in
geophysical optics and remote sensing)
Includes bibliographical references.
ISBN 0-937194-36-0
1. Economic conversion.   2. Defense industries.
3. Disarmament.   4. Nuclear nonproliferation.

Green, Alex Edward Samuel, 1919- (Cont'd)
Green, Alex Edward Samuel, 1919-   (Cont'd)
  5. World politics--1989-   I. Title. II. Series.
HC79.D4G73 1995
338.9--dc20   95-22514

Ham, Peter van, 1963-
Ham, Peter van, 1963-
    Managing non-proliferation regimes in the 1990s :
power, politics, and policies / Peter van Ham. -- New
York : Council on Foreign Relations, c1994.
    112 p. ; 22 cm.
    Includes bibliographical references (p. 100-112).
    ISBN 0-87609-161-3 : $14.95
    1. Nuclear nonproliferation. 2. World
politics--1989-   I. Title.
JX1974.73.H36 1994
327.1'74--dc20   93-38875

Han, Yong-Sup.
Han, Yong-Sup.
    Nuclear disarmament and non-proliferation in
Northeast Asia / Yong-Sup Han. -- New York : United
Nations, 1995.
    vi, 83 p. ; 21 cm. -- (Research paper / United
Nations Institute for Disarmament Research, ISSN
1014-4013 ; no. 33)
    Includes bibliographical references.
    "United Nations publication sales no. GV.E.95.0.3"-
-T.p. verso.
    "UNIDIR/95/12."
    ISBN 9290451009
    1. Nuclear disarmament--East Asia. 2. Nuclear
nonproliferation. I. United Nations Institute for
Disarmament Research. II. Title. III. Series:
Research paper (United Nations Institute for
Disarmament Research) ; no. 33.
JZ5665.H36 1995
327.1'747'095--dc21   96-216397

How Western European nuclear policy is made :
    How Western European nuclear policy is made : deciding
on the atom / edited by Harald Müller. -- New York :
St. Martin's Press, 1991.
    xiv, 241 p. : ill. ; 23 cm.
    Includes bibliographical references and index.
    ISBN 0-312-05354-1
    1. Nuclear weapons--Government policy--Europe.
    2. Nuclear energy--Government policy--Europe.
    3. Nuclear nonproliferation.   I. Müller, Harald, 1949
May 13-
U264.5.E85H68 1991
355.8'25119--dc20   90-44834
                                                    r92

Imai, Ryūkichi, 1929-
Imai, Ryūkichi, 1929-
 The long shadow of nuclear weapons : will nuclear
 electricity generation pass beyond the ten percent
 line in the long-term world primary energy supply? :
 paper prepared for the Fifteenth Oxford Energy
 Seminar, September, 1993 / Ryukichi Imai. -- Tokyo,
 Japan : International Institute for Global Peace,
 c1993.
  24 p. : ill. ; 30 cm. -- (IIGP policy paper ; 114E)
  "July 1993."
  Includes bibliographical references (p. 23-24).
  1. Nuclear industry. 2. Nuclear power plants.
 3. Nuclear weapons. 4. Nuclear nonproliferation.
 I. Oxford Energy Seminar (15th : 1993 : Tokyo, Japan)
 II. Title. III. Series.
 HD9698.A2I433   1993   94-197186

Imai, Ryūkichi, 1929-
Imai, Ryūkichi, 1929-
 Review of new mechanisms to stem nuclear
 proliferation / by Ryukichi Imai. -- Tokyo :
 International Institute for Global Peace, c1992.
  25 p. : ill. ; 30 cm. -- (IIGP policy paper ; 101E)
  "October 1992."
  Includes bibliographical references.
  1. Nuclear nonproliferation. 2. Nuclear weapons.
 I. Sekai Heiwa Kenkyūjo (Japan) II. Title.
 III. Series.
 JX1974.73.I445   1992
 327.1'74--dc20   93-233861

International cooperation on nonproliferation export controls :
 International cooperation on nonproliferation export
 controls : prospects for the 1990s and beyond / edited
 by Gary K. Bertsch, Richard T. Cupitt, and Steven
 Elliott-Gower. -- Ann Arbor : University of Michigan
 Press, c1994.
  viii, 331 p. : ill. ; 24 cm.
  Includes bibliographical references and index.
  ISBN 0-472-10515-9 (acid-free paper)
  1. Nuclear nonproliferation. 2. Export controls.
 3. International cooperation. I. Bertsch, Gary K.
 II. Cupitt, Richard T. III. Elliott-Gower, Steven.
 JX1974.73.I584   1994
 327.1'74--dc20   94-4256

International nuclear trade and nonproliferation :
 International nuclear trade and nonproliferation : the
 challenge of the emerging suppliers / edited by
 William C. Potter. -- Lexington, Mass. : Lexington
 Books, c1990.
  x, 431 p. ; 24 cm.
  "First drafts of most of the book chapters were

International nuclear trade and ... (Cont'd)
International nuclear trade and ... (Cont'd)
presented at a Conference on the Emerging Nuclear
Suppliers and Nonproliferation held at the Rockefeller
Foundation Bellagio Study and Conference Center, Villa
Serbelloni, June 22-26, 1987 [and] ... sponsored by
CISA"--Pref.
ISBN 0-669-21120-6 (alk. paper)
1. Nuclear nonproliferation--Congresses. 2. Nuclear
weapons industry--Congresses. I. Potter, William C.
II. Conference on the Emerging Nuclear Suppliers and
Nonproliferation (1987 : Rockefeller Foundation
Bellagio Study and Conference Center) III. University
of California, Los Angeles. Center for International
and Strategic Affairs.
HD9698.A2I64 1990
327.1'74--dc20   89-35601

International Seminar on Nuclear War and Planetary Emergencies (16th : 1992 : Erice, Italy)
International Seminar on Nuclear War and Planetary
Emergencies (16th : 1992 : Erice, Italy)
International Seminar on Nuclear War and Planetary
Emergencies, 16th session : proliferation of weapons
for mass destruction and cooperation on defence
systems, 2nd seminar after Rio / E. Majorana Centre
for Scientific Culture, Erice, Italy, 19-24 August
1992 ; edited by K. Goebel. -- Singapore ; River Edge,
N.J. : World Scientific, c1993.
viii, 308 p. : ill., maps ; 23 cm. -- (Science and
culture series. Nuclear strategy and peace technology)
Includes bibliographical references.
ISBN 9810214936
1. Arms race--Congresses. 2. Nuclear
nonproliferation--International cooperation-
-Congresses. 3. Weapons of mass destruction-
-Congresses. I. Goebel, K. II. Title. III. Title:
Proliferation of weapons for mass destruction and
cooperation on defence systems. IV. Series: Science
and culture series (Singapore). Nuclear strategy and
peace technology.
UA10.I59 1992
327.1'74--dc20   94-163272
                                                      r963

Israel and the non-proliferation treaty.
Israel and the non-proliferation treaty. -- [Jerusalem]
: Hebrew University of Jersualem, Leonard Davis
Institute for International Relations, [1995]
14 p. ; 30 cm. -- (Viewpoint ; no. 10)
"March 1995."
1. Nuclear weapons--Government policy--Israel.
2. Nuclear nonproliferation. 3. Treaty on the Non
-proliferation of Nuclear Weapons (1968) 4. Israel-
-Foreign relations. I. Makhon li-yeḥasim
benle'umiyim 'al-shem Le'onard Daiyis. II. Series:
Viewpoint (Makhon li-yeḥasim benle'umiyim 'al-shem

Israel and the non-proliferation treaty. (Cont'd)
Israel and the non-proliferation treaty. (Cont'd)
Le?onard Daiyis) ; no. 10.
U264.5.I75I85 1995   96-159613
                                                  r972

Jalonen, Olli-Pekka.
Jalonen, Olli-Pekka.
   Captors of denuclearization? : Belarus, Kazakhstan,
Ukraine, and nuclear disarmament / Olli-Pekka Jalonen.
-- Tampere : Tampere Peace Research Institute, 1994.
   91 p. ; 21 cm. -- (Tutkimuksia / Rauhan- ja
konfliktintutkimuslaitos, ISSN 0355-5550 ; no 54,
1994)
   Includes bibliographical references (p. 81-87).
   ISBN 9517061323
   1. Nuclear nonproliferation. 2. Arms control-
-Former Soviet republics. I. Title. II. Series:
Tutkimuksia (Rauhan- ja konfliktintutkimuslaitos
(Tampere, Finland) ; no. 54.
JX1974.73.J35   1994
327.1'74'0947--dc20   94-233485

Japan's nuclear future :
Japan's nuclear future : the plutonium debate and East
Asian security / edited by Selig S. Harrison. --
Washington, DC : Carnegie Endowment for International
Peace, c1996.
   vi, 120 p. : ill. ; 23 cm.
   "A Carnegie Endowment book."
   Includes bibliographical references.
   ISBN 0-87003-065-5
   1. Nuclear nonproliferation. 2. Nuclear weapons-
-Government policy--Japan. 3. Plutonium--Government
policy--Japan. 4. Japan--Foreign relations. 5. East
Asia--National security. I. Harrison, Selig S.
JX1974.73.J37   1996
327.1'74--dc20   96-25602

Johnson, Rebecca.
Johnson, Rebecca.
   Strengthening the non-proliferation regime : ends
and beginnings : a review of the first 1995 session of
negotiations at the Conference on Disarmament and an
Assessment of Prospects for the NPT / [written by
Rebecca Johnson with input from Martin Butcher ... [et
al.]]. -- London : British American Security
Information Council, 1995.
   56 p. ; 30 cm. -- (Acronym ; no. 6)
   Cover title.
   ISBN 1-874533-18-0
   1. Nuclear nonproliferation--Congresses. 2. Nuclear
nonproliferation--Congresses. [1. Conference on
Disarmament (United Nations) 2. Treaty on the Non
-Proliferation of Nuclear Weapons (1968)--Congresses.]
I. ACRONYM Consortium. II. Title. III. Series.

Johnson, Rebecca. (Cont'd)
Johnson, Rebecca. (Cont'd)
JX1974.73.J64 1995 96-142259

Kapur, K. D.
Kapur, K. D.
Nuclear non-proliferation diplomacy : nuclear power programmes in the Third World / K.D. Kapur. -- New Delhi, India : Lancers Books, 1993.
xv, 394 p. ; 23 cm.
I-E-70891
Includes bibliographical references and index.
ISBN 8170950368 : Rs380.00
1. Nuclear nonproliferation. 2. Nuclear industry--Developing countries. I. Title.
JX1974.73.K365 1993
327.1'74--dc20 93-900127

Karp, Aaron.
Karp, Aaron.
The United States and the Soviet Union and the control of ballistic missile proliferation to the Middle East / by Aaron Karp. -- New York : Institute for East-West Security Studies, 1989 [i.e. 1990]
32 p. ; 23 cm. -- (Public policy papers)
Includes bibliographical references (p. 32).
ISBN 0-913449-20-2 : $7.95
1. Nuclear nonproliferation--International cooperation. 2. Nuclear weapons--Middle East.
3. United States--Military policy. 4. Soviet Union--Military policy. I. Title. II. Series: Public policy papers (Institute for East-West Security Studies)
JX1974.73.K37 1990
320 s--dc20
[327.1'74] 90-44862
r91

Kelly, Peter.
Kelly, Peter.
Safeguards in Europe. -- [Vienna] : International Atomic Energy Agency, [1985]
48 p. : ill. ; 25 cm.
"Written for the IAEA by Peter Kelly"--P. 3.
1. Nuclear energy--Security measures--Europe.
2. Nuclear nonproliferation. I. International Atomic Energy Agency. II. Title.
TK9152.K45 1985
363.1'79--dc19 86-173748
r94

Kernwaffenverbreitung und internationaler Systemwandel :
Kernwaffenverbreitung und internationaler Systemwandel : neue Risiken und Gestaltungsmöglichkeiten / Joachim Krause (Hrsg.). -- 1. Aufl. -- Baden-Baden : Nomos,

Kernwaffenverbreitung und internationaler ... (Cont'd)
Kernwaffenverbreitung und internationaler ... (Cont'd)
1994.
540 p. : ill. ; 23 cm. -- (Internationale Politik und Sicherheit ; Bd. 36)
English and German.
ISBN 3-7890-3094-5
1. Nuclear nonproliferation. 2. Nuclear arms control. 3. Nuclear weapons. I. Krause, Joachim, 1951- . II. Series.
JX1974.73.K47   1994   95-180251

Kötter, Wolfgang, 1950-
Kötter, Wolfgang, 1950-
[Deutschland und die Kernwaffen. English]
Germany and the bomb : nuclear policies in the two German states, and the united Germany's nonproliferation commitments / Wolfgang Kötter, Harald Müller. -- Frankfurt am Main : Peace Research Institute Frankfurt, [1990]
iii, 40 p. ; 30 cm. -- (PRIF reports ; no. 14)
"September 1990."
Translation of: Deutschland und die Kernwaffen.
ISBN 3-926197-78-1
1. Nuclear nonproliferation. 2. Germany--Military policy. 3. Germany (West)--Military policy.
4. Germany (East)--Military policy. I. Müller, Harald, 1949 May 13- II. Title. III. Series.
JX1974.73.K6813   1990
327.1'74'0943--dc20   91-116496
                                               r96

Kötter, Wolfgang, 1950-
Kötter, Wolfgang, 1950-
Germany, Europe & nuclear non-proliferation / Wolfgang Kötter & Harald Müller. -- Southampton : Mountbatten Centre for International Studies, University of Southampton, on behalf of the Programme for Promoting Nuclear Non-Proliferation, 1991.
ii, 20 p. ; 21 cm. -- (PPNN study ; no. 1)
Cover title.
At head of title: Programme for Promoting Nuclear Non-Proliferation.
Contents: Europe and non-proliferation / by Harald Müller -- German non-proliferation policy / by Wolfgang Kötter.
ISBN 0-85432-408-9
1. Nuclear nonproliferation. 2. Nuclear arms control--Europe. 3. Germany--Military policy.
4. Europe--Military policy. I. Müller, Harald, 1949 May 13- II. Programme for Promoting Nuclear Non-Proliferation. III. Title. IV. Series.
JX1974.73.K68   1991   95-217911

Krass, Allan S.
Krass, Allan S.
  The costs, risks, and benefits of arms control /
Allan S. Krass. -- Stanford, CA : Center for
International Security and Arms Control, Stanford
University, 1996.
  54 p. : ill. ; 28 cm.
  Includes bibliographical references (p. 43-51).
  ISBN 0-935371-40-0
    1. Arms control--Verification--Costs. 2. Arms
control--Verification. 3. Nuclear arms control-
-Verification--Costs. 4. Nuclear nonproliferation.
I. Stanford University. Center for International
Security and Arms Control. II. Title.
  UA12.5.K7297   1996
  327.1'74--dc20   96-165573

Lamm, Vanda.
Lamm, Vanda.
  The utilization of nuclear energy and international
law / by Vanda Lamm ; [translated by Sándor Simon]. --
Budapest : Akadémiai Kiadó, 1984.
  155 p. ; 25 cm.
  Translated from the Hungarian.
  Includes bibliographical references and index.
  ISBN 963053892X
    1. Nuclear energy--Law and legislation. 2. Nuclear
nonproliferation.   I. Title.
  K3990.4.L3613   1984
  341.7'34--dc19   85-119666

Larkin, Bruce D., 1936-
Larkin, Bruce D., 1936-
  Nuclear designs : Great Britain, France, and China
in the global governance of nuclear arms / Bruce D.
Larkin. -- New Brunswick (U.S.A.) : Transaction,
c1996.
  354 p. ; 23 cm.
  Includes bibliographical references and index.
  ISBN 1-56000-239-5 (alk. paper)
    1. Nuclear weapons--Government policy--Great
Britain. 2. Nuclear weapons--Government policy-
-France. 3. Nuclear weapons--Government policy-
-China. 4. Nuclear arms control. 5. Nuclear
nonproliferation. 6. World politics--1989-
I. Title.
  U264.5.G7L37   1996
  327.1'74--dc20   95-23279

Leaver, Richard.
Leaver, Richard.
  The Nukem scandal and Australian safeguards / by
Richard Leaver. -- Canberra : Research School of
Pacific Studies, Australian National University, Peace

Leaver, Richard. (Cont'd)
Leaver, Richard. (Cont'd)
Research Centre, [1988]
69 p. ; 30 cm. -- (Working paper, ISSN 0817-1831 ;
no. 49)
"September 1988."
Includes bibliographical references.
ISBN 0-7315-0452-6
1. Nukem GmbH. 2. Nuclear industry--Corrupt
practices--Germany (West) 3. Nuclear fuels--Security
measures--Australia. 4. Nuclear nonproliferation-
-Government policy--Australia. I. Title.
II. Series. III. Series: Working paper (Australian
National University. Peace Research Centre) ; no. 49.
HD9698.G44N835 1988   90-118361
                                                  r94

Lee, Rensselaer W., 1937-
Lee, Rensselaer W., 1937-
    Smuggling Armageddon : the nuclear black market in
the Former Soviet Union and Europe / by Rensselaer W.
Lee. -- New York : St. Martin's Press, 1998.
    p.   cm.
    Includes bibliographical references (p.) and index.
    ISBN 0-312-21156-2
    1. Nuclear terrorism--Former Soviet republics-
-Prevention. 2. Nuclear nonproliferation. 3. Illegal
arms transfers--Former Soviet republics. 4. Black
market--Former Soviet republics. 5. Smuggling--Former
Soviet republics. 6. Nuclear terrorism--Europe,
Central--Prevention. 7. Illegal arms transfers-
-Europe, Central. 8. Black market--Europe, Central.
9. Smuggling--Europe, Central. I. Title.
HV6433.F6L44   1998
363.3'2--dc21   98-3793
                                              CIP 8/98

Limiting nuclear proliferation /
    Limiting nuclear proliferation / edited by Jed C. Snyder
and Samuel F. Wells, Jr. -- Cambridge, Mass. :
Ballinger Pub. Co., c1985.
    xxxvii, 363 p. ; 24 cm.
    Includes bibliographies and index.
    ISBN 0-88730-042-1
    1. Nuclear nonproliferation. I. Snyder, Jed C.
II. Wells, Samuel F.
JX1974.73.L56   1985
327.1'74--dc19   85-3967

Marom, Ran.
Marom, Ran.
    Israel's position on non-proliferation / by Ran
Marom. -- [Jerusalem] : Hebrew University of
Jerusalem, Leonard Davis Institute for International
Relations, [1986]
    81 p. ; 24 cm. -- (Policy studies ; 16)
    "June 1986."

Marom, Ran. (Cont'd)
Marom, Ran. (Cont'd)
Bibliography: p. 68-81.
1. Nuclear nonproliferation. 2. Nuclear disarmament--Israel. I. Title. II. Series: Policy studies (Makhon li-yeḥasim benle'umiyim ʻal-shem Le'onard Daiyis) ; 16.
JX1974.73.M375   1986
327.5694--dc19
                                     87-400278
                                            r89

Martel, William C.
Martel, William C.
Nuclear coexistence : rethinking U.S. policy to promote stability in an era of proliferation / William C. Martel, William T. Pendley. -- Montgomery, Ala. : Air University, Maxwell Air Force Base; Washington, D.C. : For sale by the Supt. of Docs., U.S. G.P.O., 1994.
vii, 178 p. ; 26 cm. -- (Air War College studies in national security ; no. 1)
Includes bibliographical references.
1. United States--Foreign relations--1993-
2. Nuclear weapons--Government policy--United States.
3. Nuclear nonproliferation. 4. National security--United States. I. Pendley, William T., 1936-   . II. Title. III. Series.
E885.M376   1994
327.73--dc20
                                     94-7940
                                         r95

Masker, John Scott.
Masker, John Scott.
Small states and security regimes : the international politics of nuclear non-proliferation in Nordic Europe and the South Pacific / John Scott Masker. -- Lanham, Md. : University Press of America, c1995.
x, 162 p. : maps ; 22 cm.
Includes bibliographical references (p. [143]-158) and index.
ISBN 0-8191-9846-3 (alk. paper)
1. Nuclear nonproliferation. 2. Nuclear-weapon-free zones--Oceania. 3. Nuclear-weapon-free zones--Scandinavia. 4. Security, International. I. Title.
JX1974.73.M377   1995
327.1'74--dc20
                                     94-46781

Mazarr, Michael J., 1965-
Mazarr, Michael J., 1965-
North Korea and the bomb : a case study in nonproliferation / Michael J. Mazarr. -- 1st ed. -- New York : St. Martin's Press, 1995.
xi, 290 p. ; 22 cm.
Includes bibliographical references (p. [241]-284) and index.
ISBN 0-312-12443-0

Mazarr, Michael J., 1965- (Cont'd)
Mazarr, Michael J., 1965- (Cont'd)
1. Nuclear weapons--Korea (North) 2. Korea (North)--Politics and government. 3. United States--Military policy. 4. Nuclear nonproliferation. I. Title.
UA853.K6M39 1995
327.1'74--dc20 94-34868

Miatello, Angelo.
Miatello, Angelo.
International nuclear agreements : a quadrilingual glossary / Angelo Miatello, Roberto Severino ; with an essay on U.S. nuclear exports and non-proliferation policy by Lawrence Scheinman and Joseph Pilat. -- Berne ; New York : P. Lang, c1988.
327, v, 59 p. ; 24 cm.
English, French, German, and Italian; prefatory material in English.
Essay has separate t.p., with title: Toward a more reliable supply : U.S. nuclear exports and non-proliferation policy / by Lawrence Sheinman and Joseph Pilat.
Essay issued: January 1986.
Includes bibliographical references.
ISBN 3-261-03888-8
1. Nuclear nonproliferation--Dictionaries--Polyglot.
2. Dictionaries, Polyglot. I. Severino, Roberto.
II. Scheinman, Lawrence. III. Pilat, Joseph F.
IV. Scheinman, Lawrence. Toward a more reliable supply. 1986. V. Title. VI. Title: Toward a more reliable supply.
JX1974.73.M527 1988
341.73403 90-105125

Molander, Roger C.
Molander, Roger C.
The nuclear asymptote : on containing nuclear proliferation / Roger C. Molander, Peter A. Wilson. -- Santa Monica, CA : Rand, 1993.
xxiii, 84 p. : ill. ; 23 cm.
"Project on Avoiding nuclear war: managing conflict in the nuclear age."
Project conducted by: Rand/UCLA Center for Soviet Studies.
ISBN 0-8330-1435-8
1. Nuclear nonproliferation. I. Wilson, Peter (Peter A.), 1943- . II. Rand/UCLA Center for Soviet Studies. III. Title.
JX1974.73.M65 1993
327.1'74--dc20 93-28969

Morrison, Philip.
Morrison, Philip.
Reason enough to hope : America and the world of the 21st century / Philip Morrison and Kosta Tsipis. -- Cambridge, Mass. : MIT Press, c1998.
p. cm.

Morrison, Philip. (Cont'd)
   Morrison, Philip. (Cont'd)
      Includes bibliographical references and index.
      ISBN 0-262-13344-X (hc : alk. paper)
      1. Nuclear nonproliferation. 2. Arms control.
   3. Security, International. 4. Nuclear arms control-
   -United States.   I. Tsipis, Kosta.   II. Title.
      JZ5675.M67   1998
      327.1'745--dc21   98-23561
                                                CIP 11/98

Mozley, Robert Fred, 1917-
   Mozley, Robert Fred, 1917-
      The politics and technology of nuclear proliferation
   / Robert F. Mozley. -- Seattle, WA : University of
   Washington Press, 1998.
         p.   cm.
      Includes bibliographical references and index.
      ISBN 0-295-97725-6 (alk. paper). -- ISBN
   0-295-97726-4 (pbk. : alk. paper)
      1. Nuclear nonproliferation.   I. Title.
      JZ5675.M69   1998
      327.1'747--dc21   98-12070
                                                CIP 9/98

Müller, Harald, 1949 May 13-
   Müller, Harald, 1949 May 13-
      After the scandals : West German nonproliferation
   policy / Harald Müller. -- Frankfurt am Main : Peace
   Research Institute Frankfurt, c1990.
      iv, 47 p. ; 30 cm. -- (PRIF reports ; no. 9)
      "February 1990."
      Includes bibliographical references (p. 42-46).
      ISBN 3-926197-64-1
      1. Nuclear nonproliferation. 2. Nuclear arms
   control--Germany (West)   I. Title. II. Series.
      JX1974.73.M84   1990   90-179933

Müller, Harald, 1949 May 13-
   Müller, Harald, 1949 May 13-
      Nuclear non-proliferation and global order / Harald
   Müller, David Fischer, and Wolfgang Kötter. -- Oxford
   ; New York : Oxford University Press ; Solna, Sweden :
   SIPRI, 1994
      xii, 258 p. : 24 cm.
      Includes bibliographical references and index.
      ISBN 0-19-829155-8 : £20.00
      1. Nuclear nonproliferation. 2. World
   politics--1989-   I. Fischer, David, 1920-
   II. Kötter, Wolfgang, 1950-   . III. Title.
      JX1974.73.M845   1994
      327.1'74--dc20   93-46769

Muller, M. Statius.
Muller, M. Statius.
The hazards arising out of the peaceful use of nuclear energy : selective bibliography / prepared by M. Statius Muller = Les risques résultant de l'utilisation pacifique de l'énergie nucléaire / bibliographie sélective préparée par M. Statius Muller. -- The Hague : Peace Palace Library, 1993.
viii, 117 p. ; 30 cm.
At head of title: Center for Studies and Research of the Hague Academy of International Law = Centre d'étude et de recherche de l'Academie de droit international de La Haye.
English and French.
Compiled exclusively from materials available in the Peace Palace Library.
Added title page title: Risques résultant de l'utilisation pacifique de l'énergie nucléaire.
1. Nuclear nonproliferation--Bibliography.
2. Nuclear energy--International cooperation--Bibliography. I. Peace Palace Library (Hague, Netherlands) II. Hague Academy of International Law. Center for Studies and Research. III. Title. IV. Title: Risques résultant de l'utilisation pacifique de l'énergie nucléaire
Z6464.D6M85 1993
[JX1974.73]
016.3271'74--dc20   94-135831
                                                r982

Mussington, David, 1960-
Mussington, David, 1960-
Arms unbound : the globalization of defense production / David Mussington. -- Washington : Brassey's, c1994.
xi, 88 p. ; 26 cm. -- (CSIA studies in international security ; no. 4)
Includes bibliographical references.
ISBN 0-02-881089-9 (pbk.)
1. Defense industries. 2. Arms transfers. 3. National security. 4. Nuclear nonproliferation. I. Title. II. Series.
HD9743.A2M87 1994
338.4'76233--dc20   94-23172

New nuclear nations :
New nuclear nations : consequences for U.S. policy / edited by Robert D. Blackwill and Albert Carnesale. -- New York : Council on Foreign Relations Press, c1993.
viii, 272 p. ; 23 cm.
"A Council on Foreign Relations book"--Cover.
Includes bibliographical references and index.
ISBN 0-87609-153-2 : $17.95
1. Nuclear nonproliferation. 2. United States--Foreign relations--1989- 3. Security, International.

New nuclear nations : (Cont'd)
New nuclear nations : consequences for U.S. ... (Cont'd)
I. Blackwill, Robert D. II. Carnesale, Albert.
JX1974.73.N477   1993
327.1'74--dc20
93-27807
r94

New threats :
New threats : responding to the proliferation of nuclear, chemical, and delivery capabilities in the Third World. -- [Queenstown, Md.] : Aspen Strategy Group ; Lanham, Md. : University Press of America, 1990.
   xi, 273 p. : ill. ; 24 cm.
   Includes bibliographical references.
   ISBN 0-8191-7670-2 (alk. paper). -- ISBN 0-8191-7671-0 (pbk. : alk. paper)
   1. Nuclear nonproliferation. 2. Chemical warfare (International law) 3. Developing countries--Foreign relations. I. Aspen Strategy Group (U.S.)
JX1974.73.N48   1990
327.1'74--dc20
89-29096

Non-proliferation :
Non-proliferation : the why and the wherefore / edited by Jozef Goldblat. -- London ; Philadelphia : Taylor & Francis, 1985.
   xi, 343 p. ; 24 cm.
   "Stockholm International Peace Research Institute."
   Includes bibliographies.
   ISBN 0-85066-304-0 : £25.00
   1. Nuclear nonproliferation. I. Goldblat, Jozef.
II. Stockholm International Peace Research Institute.
JX1974.73.N65   1985
327.1'74--dc19
85-230237

Non-proliferation in a changing world :
Non-proliferation in a changing world : India's policy and options / edited by Arun Kumar Banerji. -- Calcutta : Allied Publishers in collaboration with School of International Relations and Strategic Studies, Jadavpur University, 1997.
   xv, 163 p. ; 22 cm.
   I-E-97-905320; 63-92
   Summary: Papers originally presented at a seminar held in March 1995.
   Includes bibliographical references.
   ISBN 8170236398
   1. Nuclear nonproliferation--Congresses.
2. National security--India--Congresses. I. Banerji, Arun Kumar. II. Jadavpur University. School of International Relations and Strategic Studies.
JZ5675.N66   1997
327.1'747'0954--dc21
97-905320

Nonproliferation =
Nonproliferation = Nicht-Weiterverbreitung von
Atomwaffen / Dieter Deiseroth, Stig Gustafsson
(Hrsg./ed.). -- Frankfurt am Main : Haag + Herchen,
c1993.
    ii, 203 p. ; 21 cm. -- (IALANA-Schriftenreihe ; Bd.
4)
    English and German.
    "'Saving NTP and abolishing nuclear weapons', IALANA
-Colloquium in Stockholm organized by Svenska Jurister
Mot Kärnvapen, IALANA-Sektion BR Deutschland."
    ISBN 3-86137-093-X
    1. Nuclear nonproliferation. 2. Nuclear arms
control. I. Deiseroth, Dieter. II. Gustafsson,
Stig. III. Svenska jurister mot kärnvapen.
IV. International Association of Lawyers Against
Nuclear Arms. Sektion Bundesrepublik Deutschland.
V. Title: Nicht-Weiterverbreitung von Atomwaffen.
VI. Series.
JX1974.73.N64   1993   94-217984

**Nonproliferation and arms control assessment of weapons
-usable fissile material storage and excess plutonium
disposition alternatives.**
Nonproliferation and arms control assessment of weapons
-usable fissile material storage and excess plutonium
disposition alternatives. -- [Washington, DC] : U.S.
Dept. of Energy, Office of Arms Control and
Nonproliferation, 1997.
    xxv, 232 ; ill.; 28 cm.
    At head of title: The United States Department of
Energy.
    "DOE/NN-0007"--Cover.
    1. Plutonium--Storage--Government policy--United
States. 2. Nuclear weapons plants--United States-
-Security measures. 3. Radioactive waste disposal-
-United States. 4. Nuclear nonproliferation.
I. United States. Dept. of Energy.
TK9152.165.N66   1997
355.6'213--dc21   97-152402

The Nonproliferation predicament /
The Nonproliferation predicament / edited by Joseph F.
Pilat. -- New Brunswick, N.J. : Transaction Books,
c1985.
    ix, 137 p. ; 24 cm.
    Includes bibliographies.
    ISBN 0-88738-047-6
    1. Nuclear nonproliferation. I. Pilat, Joseph F.
JX1974.73.N67   1985
327.1'74--dc19   85-975
                                                               r90

Nonproliferation primer :
  Nonproliferation primer : preventing the spread of
  nuclear, chemical, and biological weapons / Randall
  Forsberg ... [et al.]. -- Cambridge, Mass. : MIT
  Press, 1995.
    x, 149 p. : ill. ; 24 cm.
    Includes bibliographical references and index.
    ISBN 0-262-06183-X (hc : acid-free). -- ISBN
  0-262-56095-X (pb : acid-free)
    1. Nuclear nonproliferation. 2. Biological arms
  control. 3. Chemical arms control. I. Forsberg,
  Randall.
  JX1974.73.N68   1995
  327.1'74--dc20   95-11840

Nonproliferation regimes :
  Nonproliferation regimes : policies to control the
  spread of nuclear, chemical, and biological weapons
  and missiles : report / prepared for the Committee on
  Foreign Affairs, U.S. House of Representatives, by the
  Congressional Research Service, Library of Congress. -
  - Washington : U.S. G.P.O. : For sale by the U.S.
  G.P.O., Supt. of Docs., Congressional Sales Office,
  1993.
    v, 74 p. ; 24 cm.
    At head of title: 103d Congress, 1st session.
    Distributed to some depository libraries in
  microfiche.
    Shipping list no.: 93-0202-P.
    "March 1993."
    Includes bibliographical references.
    Supt. of Docs. no.: Y 4.F 76/1:N 73/3
    GPO: 1017-A
    GPO: 1017-B (MF)
    ISBN 0-16-040282-4
    1. Nuclear nonproliferation. 2. Nuclear arms
  control--United States. 3. Chemical arms control.
  4. Biological arms control. 5. United States-
  -Military policy. I. United States. Congress.
  House. Committee on Foreign Affairs. II. Library of
  Congress. Congressional Research Service.
  JX1974.73.N643   1993
  327.1'74--dc20
  DGPO/DLC
    for Library of Congress   93-230952

Nuclear energy and security in the former Soviet Union /
  Nuclear energy and security in the former Soviet Union /
  edited by David R. Marples and Marilyn J. Young. --
  Boulder, Colo : Westview Press, 1997.
    177 p. ; 24 cm.
    Includes bibliographical references and index.
    ISBN 0-8133-9013-3 (alk. paper)
    1. Nuclear industry--Former Soviet republics.
  2. Nuclear industry--Security measures--Former Soviet

Nuclear energy and security in the former ... (Cont'd)
Nuclear energy and security in the former ... (Cont'd) republics. 3. Nuclear industry--Former Soviet republics--Safety measures. 4. Nuclear nonproliferation. I. Marples, David R. II. Young, Marilyn J., 1942- .
HD9698.F62N78 1997
363.17'99'0947--dc21 97-14352

Nuclear export controls in Europe /
Nuclear export controls in Europe / Harald Müller (ed.). -- Brussels : European Interuniversity Press, [1995?] 275 p. : ill. ; 25 cm. -- (Collection "La cité européenne" ; no 6)
"Peace Research Institute Frankfurt"--Cover.
Includes bibliographical references.
ISBN 9052014132
1. Nuclear nonproliferation. 2. Export controls--Europe. I. Müller, Harald, 1949 May 13- II. Hessische Stiftung Friedens- und Konfliktforschung. III. Series.
JX1974.73.N746 1995
327.1'74--dc20 96-123722

Nuclear non-proliferation :
Nuclear non-proliferation : an agenda for the 1990s / edited by John Simpson. -- Cambridge [Cambridgeshire] ; New York : Cambridge University Press, 1987.
xvi, 237 p. ; 24 cm. -- (Ford/Southampton studies in North/South security relations)
Papers from the Sarnia Symposium held on the island of Guernsey, 17-20 March, 1986.
Includes bibliographies and index.
ISBN 0-521-33308-3
1. Nuclear nonproliferation. I. Simpson, John, 1943- . II. Sarnia Symposium (1986 : Guernsey, Channel Islands) III. Series.
JX1974.73.N76 1987
327.1'74--dc19 87-9362
r883

Nuclear non-proliferation :
Nuclear non-proliferation : a reference handbook / edited and compiled by Darryl Howlett and John Simpson ; with contributions by the following, Philip Acton ... [et al.]. -- Harlow, Essex, U.K. : Longman Group, UK ; Detroit, MI, USA : Distributed exclusively in the United States and Canada by Gale Research, c1992.
x, 406 p. ; 25 cm. -- (Longman current affairs)
Includes bibliographical references (p. 395-400) and index.
ISBN 0-582-09648-0
1. Nuclear nonproliferation. I. Simpson, John, 1943- . II. Howlett, Darryl A., 1954- III. Acton, Philip. IV. Series.

Nuclear non-proliferation : (Cont'd)
Nuclear non-proliferation : a reference ... (Cont'd)
JX1974.73.N757   1992
327.1'74--dc20   93-135521

Nuclear non-proliferation, 1945-1991
Nuclear non-proliferation, 1945-1991 [microform]. --
Alexandria, Va. : Chadwyck-Healey, 1992.
448 microfiches.
"Project director, Virginia I. Foran"--Guide t.p.
"Materials were identified, assembled, obtained, and
indexed by the National Security Archive."
Accompanied by 2 vol. guide entitled: U.S. nuclear
non-proliferation policy, 1945-1991 : guide and index.
c1991.
   Guide no.: JX1974.73.U18 1991 <Ser>.
   1. Nuclear nonproliferation--History--Sources.
2. Nuclear disarmament--United States--History-
-Sources.   I. Foran, Virginia I.   II. National
Security Archive (U.S.)   III. U.S. nuclear non
-proliferation policy, 1945-1991.
[JX1974.73]
Microfiche 93/41 (J) <MicRR>
327.1'74'0873--dc20   93-629991

Nuclear non-proliferation and global security /
Nuclear non-proliferation and global security / edited
by David Dewitt. -- New York : St. Martin's Press,
1987.
   x, 283 p. ; 23 cm.
   Original drafts of chapters were presented at the
Conference on Global Security and the Future of the
Non-Proliferation Treaty: a Time for Reassessment,
York University, May 1985.
   Includes bibliographies and index.
   ISBN 0-312-00367-6 : 25.00 (est.)
   1. Nuclear nonproliferation.   2. Security,
International.   I. Dewitt, David B. (David Brian),
1948-   .   II. Conference on Global Security and the
Future of the Non-Proliferation Treaty: a Time for
Reassessment (1985 : York University)
JX1974.73.N77   1987
327.1'74--dc19   86-21004
                                                        r88

Nuclear non-proliferation and the non-proliferation treaty /
Nuclear non-proliferation and the non-proliferation
treaty / M.P. Fry, N.P. Keatinge, J. Rotblat (eds.). -
- Berlin ; New York : Springer-Verlag, c1990.
   xx, 198 p. : ill. ; 25 cm.
   Spine title: Nuclear non-proliferation.
   Proceedings of a symposium held in May 1989 in
Dublin, Ireland.
   Includes bibliographical references and index.
   ISBN 0-387-51756-1 (U.S. : alk. paper)
   1. Nuclear nonproliferation.   I. Fry, M. P.

Nuclear non-proliferation and the ... (Cont'd)
Nuclear non-proliferation and the ... (Cont'd)
(Michael P.), 1940- . II. Keatinge, Patrick, 1939-
. III. Rotblat, Joseph, 1908- . IV. Title: Nuclear
non-proliferation.
JX1974.73.N775 1990
327.1'74--dc20 89-26202

Nuclear-non-proliferation in India and Pakistan :
Nuclear-non-proliferation in India and Pakistan : South
Asian perspectives / edited by P.R. Chari, Pervaiz
Iqbal Cheema, Iftekharuzzaman. -- New Delhi : Manohar,
1996.
    236 p. ; 23 cm.
    I-E-96-904933; 63-92
    "Regional Centre for Strategic Studies, Colombo."
    Summary: Collection of essays.
    Includes bibliographical references (p. [211]-224)
and index.
    ISBN 8173041539
    1. Nuclear weapons--India. 2. Nuclear weapons-
-Pakistan. 3. Nuclear arms control--South Asia.
4. Nuclear nonproliferation. I. Char, P. R.
II. Cheema, Parvaiz Iqbal, 1940- .
III. Iftekharuzzaman. IV. Regional Centre for
Strategic Studies (Colombo, Sri Lanka)
UA840.N83 1996
327.1'747'0954--dc21 96-904933
                                            r97

Nuclear non-proliferation problems and control :
Nuclear non-proliferation problems and control : first
symposium / Nuclear Society International, Moscow
[and] Group on Problems of Nuclear Weapons Non
-Proliferation, Control of Nuclear Materials, and
Liability for Nuclear Damage. -- Moscow : The Society,
1993.
    99 p. : ill. ; 21 cm.
    Includes bibliographical references.
    1. Nuclear nonproliferation. I. Nuclear Society
International. II. Group on Problems of Nuclear
Weapons Non-Proliferation, Control of Nuclear
Materials, and Liability for Nuclear Damage.
JX1974.73.N778 1993
341.7'34--dc20 94-131660

The nuclear non-proliferation regime :
The nuclear non-proliferation regime : prospects for the
21st century / edited by Raju G.C. Thomas. --
Houndmills, Basingstoke, Hampshire : Macmillan Press ;
New York : St. Martin's Press, Scholarly and Reference
Division, 1998.
    xix, 359 p. ; 23 cm.
    Includes bibliographical references and index.
    ISBN 0-333-68964-X (cloth). -- ISBN 0-312-21042-6
(St. Martin's Press : cloth)
    1. Nuclear nonproliferation. I. Thomas, Raju G. C.

The nuclear non-proliferation regime : (Cont'd)
The nuclear non-proliferation regime : ... (Cont'd)
KZ5675.N83   1998
341.7'34--dc21   97-37889

The Nuclear Non-proliferation Treaty /
The Nuclear Non-proliferation Treaty / edited by Ian
Bellany, Coit D. Blacker, and Joseph Gallacher. --
London, England ; Totowa, N.J. : F. Cass, 1985.
134 p. : ill. ; 23 cm.
Includes bibliographical references.
Contents: The Non-proliferation Treaty as its half
-life / John Simpson -- Article I of the Non
-proliferation Treaty and United Kingdom-United States
nuclear weapon cooperation / Norman Dombey --
Assessing article II / Ladi Adenrele -- Changes in the
political and technical environments of article IV /
Philip Gummett -- Britain's performance in
implementing article VI of the NPT / Dan Keohane -- US
-Soviet strategic nuclear arms control / Steve Smith -
- Article VII the treaty of Tlatelolco and Colonial
Warfare in the 20th century / Joseph Gallacher --
Tomorrow's nuclear trading system / William Walker --
The non-proliferation of nuclear weapons / Ian
Bellany.
   ISBN 0-7146-3250-3 : $27.50 (U.S.)
   1. Nuclear nonproliferation.   I. Bellany, Ian.
II. Blacker, Coit D.   III. Gallacher, Joseph.
JX1974.73.N78   1985
341.7'34--dc19   86-121061

Nuclear policies in Northeast Asia /
Nuclear policies in Northeast Asia / edited by Andrew
Mack. -- New York ; Geneva : United Nations, 1995.
   xi, 265 p. : ill. ; 21 cm.
   At head of title: UNIDIR, United Nations Institute
for Disarmament Research, Geneva.
   "United Nations publication sales no. G.V.E.95.0.8"-
-T.p. verso.
   "From a conference ... convened by UNIDIR in Seoul,
25-27 May 1994"--P. ix.
   "UNIDIR/95/16."
   Includes bibliographical references.
   ISBN 9290451017
   1. Nuclear nonproliferation. 2. Nuclear
disarmament--East Asia. 3. Nuclear arms control--East
Asia. 4. Security, International.   I. Mack, Andrew,
1939-   . II. United Nations Institute for
Disarmament Research.
JX1974.73.N812   1995
                                                             96-125283
                                                                   r98

Nuclear proliferation :
Nuclear proliferation : opposing viewpoints / Charles P.
Cozic, book editor ; Karin L. Swisher, assistant

Nuclear proliferation : (Cont'd)
Nuclear proliferation : opposing viewpoints ... (Cont'd)
editor. -- San Diego, CA : Greenhaven Press, c1992.
234 p. : ill., map ; 22 cm. -- (Opposing viewpoints series)
Includes bibliographical references (p. 225-227) and index.
Summary: Presents differing opinions on the threat of nuclear proliferation, the need for arms control, the role of NATO, the elimination of nuclear weapons, and other related topics.
ISBN 1-56510-004-2 (pbk. : acid-free paper). -- ISBN 1-56510-005-0 (lib. : acid-free paper)
1. Nuclear nonproliferation--Juvenile literature.
[1. Nuclear nonproliferation. 2. Disarmament.]
I. Cozic, Charles P., 1957-   . II. Swisher, Karin, 1966-   . III. Series: Opposing viewpoints series (Unnumbered)
JX1974.73.N83   1992
327.1'74--dc20   92-23065
                                                          AC

Nuclear proliferation :
Nuclear proliferation : South Asia and the Middle East / edited by Robert H. Bruce. -- Perth, W.A. : Indian Ocean Centre for Peace Studies ; Canberra : Australian Institute of International Affairs, c1992.
viii, 152 p. ; 21 cm. -- (Monograph / Indian Ocean Centre for Peace Studies ; no. 2)
Based on papers presented to a seminar Changing prospects for peace in the Indian Ocean region, held January 14-16, 1991 at the University of Western Australia in Perth.
Includes bibliographical references.
ISBN 1-86342-007-X
1. Nuclear nonproliferation. 2. Nuclear arms control--South Asia. 3. Nuclear arms control--Middle East.   I. Bruce, Robert H.   II. Series: Monograph (Indian Ocean Centre for Peace Studies) ; no. 2.
JX1974.73.N84   1992
327.17'4--dc20   95-210318

Nuclear proliferation after the Cold War /
Nuclear proliferation after the Cold War / edited by Mitchell Reiss and Robert S. Litwak. -- Washington, D.C. : Woodrow Wilson Center Press ; Baltimore, Md. : Distributed by Johns Hopkins University Press, c1994.
ix, 370 p. ; 24 cm. -- (Woodrow Wilson Center special studies)
Includes bibliographical references and index.
ISBN 0-943875-64-1 (hardcover : acid-free paper). -- ISBN 0-943875-57-9 (pbk. : acid-free paper)
1. Nuclear nonproliferation.   I. Reiss, Mitchell.
II. Litwak, Robert.   III. Series.
JX1974.73.N813   1994
327.1'74--dc20   94-22717

Nuclear proliferation and international security /
Nuclear proliferation and international security /
edited by K. Subrahmanyam. -- New Delhi : Lancer
International in association with Institute for
Defence Studies and Analyses, 1985-86 [i.e. 1985]
    310 p. ; 22 cm.
    ISBN (invalid) 1-85127-058-4 : Rs150.00
    1. Nuclear nonproliferation. 2. Security,
International. I. Subrahmanyam, K. II. Institute
for Defence Studies and Analyses.
JX1974.73.N814   1985
327.1'74--dc19   85-904278

Nuclear proliferation and the legality of nuclear weapons
/
Nuclear proliferation and the legality of nuclear
weapons / edited by William M. Evan, Ved P. Nanda. --
Lanham, Md. : University Press of America, c1995.
    xii, 421 p. : ill. ; 23 cm.
    Includes bibliographical references and indexes.
    ISBN 0-7618-0088-3 (cloth : alk. paper). -- ISBN
0-7618-0089-1 (paper : alk. paper)
    1. Nuclear nonproliferation. 2. Nuclear arms
control. 3. Nuclear weapons (International law)
I. Evan, William M. II. Nanda, Ved P.
JX1974.73.N8145   1995
327.1'74--dc20   95-31935

Nuclear proliferation factbook /
Nuclear proliferation factbook / prepared for the
    Subcommittees on Arms Control, International Security,
    and Science and on International Economic Policy and
    Trade of the Committee on Foreign Affairs, U.S. House
    of Representatives and the Subcommittee on Energy,
    Nuclear Proliferation, and Federal [i.e. Government]
    Processes of the Committee on Governmental Affairs,
    U.S. Senate ; by the Environment and Natural Resources
    Policy Division, Congressional Research Service,
    Library of Congress. -- Washington : U.S. G.P.O. : For
    sale by the Supt. of Docs., U.S. G.P.O., 1985.
    x, 591 p. : ill. ; 24 cm.
    At head of title: 99th Congress, 1st session. Joint
committee print.
    Distributed to some depository libraries in
microfiche.
    "August 1985."
    Bibliography: p. 590-591.
    Item 1017-A, 1017-B (microfiche)
    Supt. of Docs. no.: Y 4.F 76/1:N 88/13
    $13.00
    1. Nuclear nonproliferation. I. United States.
Congress. House. Committee on Foreign Affairs.
Subcommittee on Arms Control, International Security,
and Science. II. United States. Congress. House.
Committee on Foreign Affairs. Subcommittee on

Nuclear proliferation factbook / (Cont'd)
Nuclear proliferation factbook / prepared ... (Cont'd)
International Economic Policy and Trade. III. United
States. Congress. Senate. Committee on Governmental
Affairs. Subcommittee on Energy, Nuclear
Proliferation, and Government Processes. IV. Library
of Congress. Environment and Natural Resources Policy
Division.
JX1974.73.N817   1985
327.1'74--dc19
DGPO/DLC
for Library of Congress   85-602596

Nuclear proliferation factbook /
Nuclear proliferation factbook / prepared for the
Committee on Governmental Affairs, United States
Senate ; by the Congressional Research Service,
Library of Congress. -- [5th ed.] -- Washington : U.S.
G.P.O. : For sale by the U.S. G.P.O., Supt. of Docs.,
Congressional Sales Office, 1995.
   xviii, 768 p. : ill., maps ; 24 cm. -- (S. prt. ;
103-111)
   Distributed to some depository libraries in
microfiche.
   Shipping list no.: 95-0087-P.
   "December 1994."
   Includes bibliographical references.
   Supt. of Docs. no.: Y 4.G 74/9:S.PRT.103-111
   GPO: 1037-B
   GPO: 1037-C (MF)
   ISBN 0-16-046780-2
   1. Nuclear nonproliferation. 2. Nuclear weapons.
3. Nuclear weapons--Government policy--United States.
4. Nuclear energy--Government policy--United States.
I. United States. Congress. Senate. Committee on
Governmental Affairs. II. Library of Congress.
Congressional Research Service. III. Title: 103rd
Congress, 2d session. Committee print. IV. Series.
JX1974.73.N817   1995
DGPO/DLC
for Library of Congress   95-144467

Nuclear proliferation in South Asia :
Nuclear proliferation in South Asia : containing the
threat : a staff report to the Committee on Foreign
Relations, United States Senate. -- Washington : U.S.
G.P.O. : For sale by the Supt. of Docs., Congressional
Sales Office, U.S. G.P.O., 1988.
   xi, 42 p. : 1 map ; 24 cm. -- (S. prt. ; 100-121)
   At head of title: 100th Congress, 2d session.
Committee print.
   Distributed to some depository libraries in
microfiche.
   Shipping list no.: 88-530-P.
   "August 1988."
   Item 1039-A, 1039-B (microfiche)
   Supt. of Docs. no.: Y 4.F 76/2:S.prt.100-121

Nuclear proliferation in South Asia : (Cont'd)
Nuclear proliferation in South Asia : ... (Cont'd)
1. Nuclear nonproliferation. 2. United States-
-Foreign relations--South Asia. 3. South Asia-
-Foreign relations--United States. I. United States.
Congress. Senate. Committee on Foreign Relations.
II. Series.
JX1974.73.N823   1988
327.1'74'0954--dc19
DGPO/DLC
for Library of Congress   88-602704
                                                  r92

Nuclear proliferation in South Asia :
Nuclear proliferation in South Asia : the prospects for
arms control / edited by Stephen Philip Cohen. --
Boulder, Colo. : Westview Press, c1991.
  xx, 377 p. : ill. ; 23 cm.
  Includes bibliographical references (p. 367-370) and
index.
  ISBN 0-8133-8159-2 (acid free paper)
  1. Nuclear arms control--South Asia. 2. Nuclear
arms control--Verification--South Asia. 3. Nuclear
arms control--India. 4. Nuclear arms control-
-Pakistan. 5. Nuclear nonproliferation. I. Cohen,
Stephen P., 1936-   .
JX1974.7.N819   1991
327.1'74--dc20   90-24634
                                                  r94

Nuclear proliferation in the 1980s :
Nuclear proliferation in the 1980s : perspectives and
proposals / edited by William H. Kincade and Christoph
Bertram. -- London : Macmillan Press, 1982.
  xiv, 272 p. ; 23 cm.
  Bibliography: p. [259]-266.
  Includes index.
  ISBN 0-333-32304-1
  1. Nuclear nonproliferation--Congresses. 2. Nuclear
disarmament--Congresses. I. Kincade, William H.
II. Bertram, Christoph, 1937-
[JX1974.73.N825x   1982b]
MH
for Library of Congress   84-672919
                                                  r90

Nuclear rivalry and international order /
Nuclear rivalry and international order / edited by Jørn
Gjelstad and Olav Njølstad. -- London ; Thousand Oaks
: SAGE Publications, 1996.
  x, 212 p. ; 25 cm.
  "PRIO, International Peace Research Institute,
Oslo"--T.p.
  "The essays in this anthology were commissioned for
a conference held in June 1993 in Rjukan, in the
county of Telemark, Norway"--Foreword.
  Includes bibliographical references and indexes.

Nuclear rivalry and international order / (Cont'd)
Nuclear rivalry and international order / ... (Cont'd)
ISBN (invalid) 0-8039-7753-0
1. Nuclear weapons--Congresses. 2. Nuclear disarmament--Congresses. 3. Nuclear nonproliferation--Congresses. 4. Security, International--Congresses. 5. Peace--Congresses. I. Gjelstad, Jørn. II. Njølstad, Olav. III. International Peace Research Institute.
U264.N8154 1996
327.1'747--dc21   96-160103

Nuclear war, nuclear proliferation and their consequences ; proceedings of the Vth international colloquium organized by the Groupe de Bellerive, Geneva, 27-29 June 1985 /
Nuclear war, nuclear proliferation and their consequences ; proceedings of the Vth international colloquium organized by the Groupe de Bellerive, Geneva, 27-29 June 1985 / edited by Sadruddin Aga Khan. -- Oxford : Clarendon Press ; New York : Oxford University Press, 1986.
xii, 483 p. : ill. ; 24 cm.
Includes index.
ISBN 0-19-825543-8 : $27.95. -- ISBN 0-19-825542-X : $13.95
1. Nuclear warfare--Congresses. 2. Nuclear weapons--Congresses. 3. Arms race--History--20th century--Congresses. 4. Nuclear nonproliferation--Congresses. 5. Nuclear arms control--Congresses. I. Aga Khan, Sadruddin, Prince, 1933- . II. Groupe de Bellerive.
U263.N7784 1986
351'.0217--dc19   85-15433

Nuclear weapons after the comprehensive test ban :
Nuclear weapons after the comprehensive test ban : implications for modernization and proliferation / edited by Eric Arnett. -- Solna, Sweden : Sipri ; New York : Oxford University Press, 1996.
x, 150 p. ; 24 cm.
Includes bibliographical references and index.
ISBN 0-19-829194-9 : £20.00 : Formerly CIP
1. Nuclear nonproliferation. 2. Nuclear arms control. 3. Nuclear weapons--Government policy. 4. Nuclear weapons--Testing. I. Arnett, Eric H.
JX1974.73.N87 1996
327.174--dc20
Uk
for Library of Congress   96-222615

Nuclear weapons, arms control, and the threat of thermonuclear war, special studies. Seventh supplement, 1993-1995
Nuclear weapons, arms control, and the threat of thermonuclear war, special studies. Seventh supplement, 1993-1995 [microform]. -- Bethesda, Md. :

Nuclear weapons, arms control, and the threat ... (Cont'd)
Nuclear weapons, arms control, and the ... (Cont'd)
University Publications of America, 1996.
  microfilm reels ; 35 mm. -- ([The special studies series])
  Accompanied by a printed guide compiled by Blair D. Hydrick.
  ISBN 1-55655-539-3
  1. Nuclear arms control. 2. Nuclear weapons.
  3. Nuclear nonproliferation. 4. Nuclear disarmament.
  5. Biological arms control. 6. Biological weapons.
  7. Chemical arms control. 8. Chemical weapons.
  I. Hydrick, Blair. II. Series: Special studies series (University Publications of America (Firm))
  [JX1974.7]
  327.1'72--dc20   96-41845
                                                   CIP 10/96

O'Heffernan, Patrick.
O'Heffernan, Patrick.
  The first nuclear world war : a strategy for preventing nuclear wars and the spread of nuclear weapons / Patrick O'Heffernan, Amory B. Lovins, L. Hunter Lovins. -- London : Hutchinson, 1984, c1983.
  444 p. : ill. ; 24 cm.
  Bibliography: p. 411-429.
  Includes index.
  ISBN 0-09-155830-1
  1. Nuclear warfare. 2. World War III. 3. Nuclear industry. 4. Nuclear nonproliferation. 5. Nuclear disarmament. 6. Energy policy. I. Lovins, Amory B., 1947- . II. Lovins, L. Hunter, 1950-
  III. Title.
  [U263.036x 1984]
  MH
  for Library of Congress   87-673196

On the brink :
On the brink : nuclear proliferation and the Third World / [edited by Peter Worsley, Kofi Buenor Hadjor]. -- London : Third World Communications, Kwame Nkrumah House, 1987.
  278 p. ; 22 cm. -- (Third World book series)
  Includes bibliographical references (p. [273]-277).
  ISBN 1-870101-50-2. -- ISBN 1-870101-55-3
  1. Nuclear nonproliferation. 2. Nuclear arms control. I. Worsley, Peter. II. Hadjor, Kofi Buenor. III. Series.
  JX1974.73.05   1987   90-180387

Opaque nuclear proliferation :
Opaque nuclear proliferation : methodological and policy implications / edited by Benjamin Frankel. -- London, England ; Portland, OR : F. Cass, 1991.
  201 p. ; 23 cm.
  First published in the Journal of strategic studies, vol. 13, no. 3.

Opaque nuclear proliferation : (Cont'd)
Opaque nuclear proliferation : ... (Cont'd)
Includes bibliographical references.
ISBN 0-7146-3418-2
1. Nuclear nonproliferation. I. Frankel, Benjamin,
1949- .
JX1974.73.O73   1991
327.1'74--dc20   91-10608
                                              r94

Paranjpe, Shrikant.
Paranjpe, Shrikant.
   US nonproliferation policy in action, South Asia /
Shrikant Paranjpe. -- New York : Envoy Press, c1987.
   viii, 142 p. ; 23 cm.
   Spine title: U.S. nonproliferation policy in action,
South Asia.
   Bibliography: p. [126]-135.
   Includes index.
   ISBN 0-938719-18-1
   1. Nuclear nonproliferation. 2. Export controls-
-United States. 3. Technology transfer--Law and
legislation--United States. 4. Nuclear energy-
-Government policy--India. 5. Nuclear weapons-
-Pakistan. I. Title. II. Title: U.S.
nonproliferation policy in action, South Asia.
JX1974.73.P37   1987b
327.1'74--dc19   87-80662

Paranjpe, Shrikant.
Paranjpe, Shrikant.
   US nonproliferation policy in action, South Asia /
Shrikant Paranjpe. -- New Delhi : Sterling Publishers,
c1987.
   viii, 142 p. ; 23 cm.
   I E 56141
   Summary: With special reference to India and
Pakistan.
   Bibliography: p. [126]-135.
   Includes index.
   Spine title: U.S. nonproliferation policy in action,
South Asia.
   ISBN 8120707257 : Rs125.00
   1. Nuclear nonproliferation. 2. Export controls-
-United States. 3. Technology transfer--Law and
legislation--United States. 4. Nuclear energy-
-Government policy--India. 5. Nuclear weapons-
-Pakistan. I. Title. II. Title: U.S.
nonproliferation policy in action, South Asia.
JX1974.73.P37   1987
327.1'74'0954--dc19   87-904890

Paul, T. V.
Paul, T. V.
   Reaching for the bomb : the Indo-Pak nuclear

Paul, T. V. (Cont'd)
Paul, T. V. (Cont'd)
scenario / T.V. Paul. -- New Delhi : Dialogue, c1984.
199 p. : ill. ; 23 cm.
I E 48978
Bibliography: p. [193]-199.
Rs60.00
1. Nuclear nonproliferation. 2. Nuclear weapons-
-India. 3. Nuclear weapons--Pakistan. I. Title.
JX1974.P385 1984
327.1'74--dc19   85-900330

Plutonium and security :
Plutonium and security : the military aspects of the
plutonium economy / edited by Frank Barnaby. -- New
York : St. Martin's Press, 1992.
xvi, 296 p. : ill. ; 23 cm.
Includes bibliographical references and index.
ISBN 0-312-06724-0
1. Reactor fuel reprocessing--United States.
2. Plutonium. 3. Nuclear industry--Military aspects-
-United States. 4. Nuclear nonproliferation.
I. Barnaby, Frank.
TK9360.P546 1991
355.8'25119--dc20   91-15993

Poulose, T. T.
Poulose, T. T.
United Nations and nuclear proliferation / T.T.
Poulose. -- Delhi : B.R. Pub. Corp., 1988.
vi, 250 p. ; 23 cm.
I E 57055
Includes bibliographies and index.
ISBN 817018472X : Rs125.00
1. Nuclear nonproliferation. 2. United Nations.
I. Title.
JX1974.73.P68 1988
327.1'74--dc19   88-900892

Proliferation and export controls /
Proliferation and export controls / edited by Kathleen
Bailey & Robert Rudney. -- Lanham : University Press
of America ; Fairfax, Va. : National Institute for
Public Policy, c1993.
xx, 117 p. ; 24 cm.
Based on papers presented at a January 1992
conference by the National Institute for Public Policy
and the Peace Research Institute of Frankfurt.
Includes bibliographical references.
ISBN 0-8191-8720-8 (paper : alk paper). -- ISBN
0-8191-8719-4 (cloth : alk paper)
1. Nuclear nonproliferation. 2. Arms control.
3. Export controls. I. Bailey, Kathleen C.
II. Rudney, Robert.
JX1974.73.P77 1993
327.1'74--dc20   92-23886

Proliferation and international security :
Proliferation and international security : converging roles of verification, confidence building, and peacekeeping / edited by Steven Mataija and Lyne C. Bourque. -- Toronto, Canada : Centre for International and Strategic Studies, York University, 1993.
xi, 217 p. : ill. ; 22 cm.
C94-930116-7
"The chapters included in this volume are based on papers commissioned for and presented at the Tenth Annual Ottawa Verification Symposium ... held in Montebello, Québec, 24-27 February 1993"--P. x-xi.
Includes bibliographical references.
ISBN 0-920231-15-2
1. Nuclear nonproliferation. 2. Arms control--Verification. 3. Security, International.
I. Mataija, Steven. II. Bourque, Lyne C., 1970-
III. York Centre for International and Strategic Studies. IV. Ottawa Verification Symposium (10th : 1993 : Montebello, Québec)
JX1974.73.P775 1993
327.1'74--dc20   94-173538

Proliferation and the Former Soviet Union.
Proliferation and the Former Soviet Union. -- Washington, DC : Office of Technology Assessment, Congress of the U.S. : For sale by the U.S. G.P.O., Supt. of Docs., [1994]
x, 92 p. : ill., maps ; 28 cm.
"September 1994"--T.p. verso.
Includes bibliographical references and index.
Shipping list no.: 94-0302-P.
S/N 052-003-01384-3 (GPO)
Supt. of Docs. no.: Y 3.T 22/2:2 P 94/2
GPO: 1070-M
ISBN 0-16-045160-4 : $6.50
1. Nuclear nonproliferation. 2. Former Soviet republics. 3. Nuclear arms control--Verification--Russia (Federation) 4. Security, International.
I. United States. Congress. Office of Technology Assessment.
U264.P76 1994
327.1'74'0947--dc20
DGPO/DLC
for Library of Congress   94-219219

Proliferation concerns :
Proliferation concerns : assessing U.S. efforts to help contain nuclear and other dangerous materials and technologies in the former Soviet Union / Office of International Affairs, National Research Council. -- Washington, D.C. : National Academy Press, 1997.
xv, 142 ; 23 cm.
Includes bibliographical references.
ISBN 0-309-05741-8

Proliferation concerns : (Cont'd)
Proliferation concerns : assessing U.S. ... (Cont'd)
1. Nuclear nonproliferation--International cooperation. 2. Technical assistance, American--Former Soviet republics. 3. Export controls--Former Soviet republics. 4. Nuclear disarmament--Former Soviet republics. 5. Weapons of mass destruction--Former Soviet republics. I. National Research Council (U.S.). Office of International Affairs.
JZ5675.P75   1997
327.1'747'0947--dc21   97-66336

Proliferation in all its aspects post-1995 :
Proliferation in all its aspects post-1995 : the verification challenge and response : symposium proceedings / edited by J. Marshall Beier and Steven Mataija. -- Toronto, Canada : Centre for International and Strategic Studies, York University, 1995.
  xii, 148 p. ; 23 cm.
  Papers presented at the 12th Annual Ottawa Verification Symposium, held in Montebello, Québec, March 8-11, 1995.
  ISBN 0-920231-20-9
  1. Arms control--Verification--Congresses.
2. Nuclear nonproliferation--Congresses. 3. Security, International--Congresses. I. Beier, J. Marshall.
II. Mataija, Steven. III. York Centre for International and Strategic Studies. IV. Ottawa Verification Symposium (12th : 1995 : Montebello, Québec)
UA12.5.P755   1995   96-205958

The proliferation puzzle :
The proliferation puzzle : why nuclear weapons spread and what results / Zachary S. Davis and Benjamin Frankel, editors. -- London, England ; Portland, OR : F. Cass, 1993.
  356 p. : ill. ; 24 cm.
  "First appeared in a special issue ... of Security studies 2, nos. 3/4 (spring/summer 1993)"--T.p. verso.
  Includes bibliographical references.
  ISBN 0-7146-4546-X. -- ISBN (invalid) 1-71464-108-1 (pbk. : $37.50)
  1. Nuclear nonproliferation. I. Davis, Zachary S., 1955- . II. Frankel, Benjamin, 1949- .
JX1974.73.P78   1993
327.1'74--dc20   93-28873

Proliferation, theater missile defense, and U.S. security.
Proliferation, theater missile defense, and U.S. security. -- Cambridge, Mass. : Institute for Foreign Policy Analysis, c1994.
  xxi, 51 p. ; 23 cm.
  "Report of an IFPA Working Group."
  ISBN 0-89549-103-6 : $9.75
  1. United States--Military policy. 2. National security--United States. 3. Nuclear nonproliferation.

Proliferation, theater missile defense, and ... (Cont'd)
Proliferation, theater missile defense, and ... (Cont'd)
I. Institute for Foreign Policy Analysis.
UA23.P76 1994
327.73--dc20   93-49046

Pugwash Conference on Science and World Affairs (38th : 1988 : Dagomys, Russia)
Pugwash Conference on Science and World Affairs (38th : 1988 : Dagomys, Russia)
  Global problems and common security : annals of Pugwash 1988 / edited by J. Rotblat and V.I. Goldanskii. -- Berlin ; New York : Springer -Verlag, c1989.
    xi, 267 p. : ill. ; 24 cm.
    Includes bibliographical references.
    ISBN 0-387-51699-9 (U.S. : alk. paper)
    1. Nuclear disarmament--Congresses. 2. Nuclear nonproliferation--Congresses. 3. Chemical warfare (International law)--Congresses. I. Rotblat, Joseph, 1908-   . II. Gol'danskiĭ, V. I. (Vitaliĭ Iosifovich) III. Title.
JX1974.7.P836 1988
327.1'74--dc20   89-21964
                                                             r94

Pugwash Conference on Science and World Affairs (44th : 1994 : Kolymparion, Greece)
Pugwash Conference on Science and World Affairs (44th : 1994 : Kolymparion, Greece)
  Towards a war-free world : annals of Pugwash 1994 / edited by Joseph Rotblat. -- Singapore ; River Edge, NJ : World Scientific, c1995.
    xi, 192 p. : ill. ; 23 cm.
    Proceedings of the 44th annual Pugwash Conference, held in July 1994 in Crete.
    Includes bibliographical references and index.
    ISBN 9810224923
    1. Nuclear nonproliferation--Congresses.
    2. Security, International--Congresses. I. Rotblat, Joseph, 1908-   . II. Title.
JX1974.73.P84 1994
327.1'74--dc20   96-114571
                                                             r97

Pulling back from the nuclear brink :
  Pulling back from the nuclear brink : reducing and countering nuclear threats / edited by Barry R. Schneider and William L. Dowdy. -- London ; Portland, OR : F. Cass, 1998.
    xiv, 309 p. : ill. ; 24 cm.
    Includes bibliographical references and index.
    ISBN 0-7146-4856-6 (cloth). -- ISBN 0-7146-4412-9 (pbk)
    1. Nuclear nonproliferation. 2. Nuclear arms control--United States. 3. United States--Military policy. I. Schneider, Barry R. II. Dowdy, William

Pulling back from the nuclear brink : (Cont'd)
Pulling back from the nuclear brink : ... (Cont'd)
L., 1944- .
JZ5675.P85 1998
327.1'747--dc21   97-53200

Quester, George H.
   Quester, George H.
      Nuclear Pakistan and nuclear India : stable
   deterrent or proliferation challenge? / George H.
   Quester. -- [Carlisle Barracks, Pa.] : Strategic
   Studies Institute, U.S. Army War College, [1992]
      iv, 22 p. ; 23 cm.
      "November 25, 1992."
      Includes bibliographical references (p. 22).
      1. Nuclear weapons--Pakistan. 2. Nuclear weapons-
   -India. 3. Pakistan--Strategic aspects. 4. India-
   -Strategic aspects. 5. Nuclear nonproliferation.
   I. Army War College (U.S.). Strategic Studies
   Institute. II. Title.
   UA853.P3Q47 1992
   327.1'74'0954--dc20   94-133139

Quester, George H.
   Quester, George H.
      Peaceful P.A.L. / by George H. Quester. -- [Los
   Angeles] : Center for Arms Control and International
   Security, University of California, Los Angeles,
   [1977]
      23 p. ; 28 cm. -- (ACIS working paper ; no. 9)
      "November 1977."
      Bibliography: p. 22-23.
      1. Nuclear nonproliferation. 2. Nuclear reactors.
   I. Title. II. Title: Peaceful PAL. III. Series: ACIS
   working paper (Los Angeles, Calif. : 1976) ; no. 9.
   [JX1974.73Q47x 1977]
   MH
      for Library of Congress   85-673451
                                                                  r89

Regional approaches to curbing nuclear proliferation in the Middle East and South Asia /
   Regional approaches to curbing nuclear proliferation in
   the Middle East and South Asia / edited by Tariq Rauf.
   -- Ottawa, Ont., Canada : Canadian Centre for Global
   Security, 1992.
      viii, 134 p. ; 28 cm. -- (Aurora papers ; 16)
      C93-90052-9
      Spine title: Regional approaches to non
   -proliferation.
      Includes bibliographical references.
      ISBN 0-920357-33-4 : $12.00
      1. Nuclear nonproliferation. 2. Nuclear arms
   control--Middle East. 3. Nuclear arms control--South
   Asia. I. Rauf, Tariq, 1949- . II. Canadian Centre
   for Global Security. III. Title: Regional approaches
   to non-proliferation. IV. Series.

Regional approaches to curbing nuclear ... (Cont'd)
Regional approaches to curbing nuclear ... (Cont'd)
JX1974.73.R438   1992
327.1'74--dc20   93-148906

Regional security and nonproliferation /
Regional security and nonproliferation / Seiitsu
Tachibana (ed.). -- 1st. ed. -- Mosbach : AFES-PRESS,
1996.
   137 p. ; 30 cm. -- (IPRA defense and disarmament
study group paper ; 7) (AFES-PRESS report ; v. 57)
   Includes bibliographical references (p. 121-134).
   ISBN 3-926979-61-5
   1. Security, International. 2. Nuclear
nonproliferation. I. Tachibana, Seiitsu.
II. Series.   III. Series: AFES-PRESS report ; Bd. 57.
JZ5588.R44   1996   96-167323

The Regulation of nuclear trade :
The Regulation of nuclear trade : non-proliferation,
supply, safety. -- Paris, France : Nuclear Energy
Agency, Organisation for Economic Co-operation and
Development ; [Washington, D.C. : OECD Publications
and Information Centre, distributor], 1988.
   2 v. : ill. (some col.) ; 27 cm. -- (Nuclear
legislation series)
   Published also in French under title: Réglementation
du commerce nucléaire.
   Bibliography: v. 1, p. 257-262.
   Contents: v. 1. International aspects -- v.
2. National regulations.
   ISBN 9264131205 (v. 1) : 270F. -- ISBN 9264131213
(v. 2) : 270F
   1. Nuclear nonproliferation. 2. Nuclear industry-
-Law and legislation. 3. Radioactive substances--Law
and legislation.   I. OECD Nuclear Energy Agency.
II. Series.
JX1974.73.R44   1988
327.1'74--dc20   88-210497
                                                           r89

Reiss, Mitchell.
   Reiss, Mitchell.
   Bridled ambition : why countries constrain their
nuclear capabilities / Mitchell Reiss. -- Washington,
D.C., U.S.A. : Woodrow Wilson Center Press ;
Baltimore, Md. : Distributed by the Johns Hopkins
University Press, c1995.
   346 p. ; 24 cm. -- (Woodrow Wilson Center special
studies)
   Includes bibliographical references and index.
   ISBN 0-943875-72-2 (cloth : acid-free paper). --
ISBN 0-943875-71-4 (pbk. : acid-free paper)
   1. Nuclear weapons--Government policy--Case studies.
2. Nuclear nonproliferation--Case studies.   I. Title.
II. Series.

Reiss, Mitchell. (Cont'd)
Reiss, Mitchell. (Cont'd)
U264.R45   1995
355.02'17--dc20   95-2646

Reiss, Mitchell.
Reiss, Mitchell.
   Without the bomb : the politics of nuclear
nonproliferation / Mitchell Reiss. -- New York :
Columbia University Press, 1988.
   xxii, 337 p. ; 24 cm.
   Revision of author's thesis (Ph. D.)--Oxford
University.
   Bibliography: p. [277]-331.
   Includes index.
   ISBN 0-231-06438-1 [alk. paper]
   1. Nuclear nonproliferation.   I. Title.
JX1974.73.R45   1988
327.1'74--dc19   87-17395

The Role of the International Atomic Energy Agency following the third review conference of the non-proliferation treaty :
   The Role of the International Atomic Energy Agency
following the third review conference of the non
-proliferation treaty : an exchange of letters. --
[London] : Foreign and Commonwealth Office, [1986]
   4, 5 p. ; 30 cm. -- (Foreign policy document ; no
136)
   "December 1985, February 1986."
   1. Nuclear nonproliferation.   2. International
Atomic Energy Agency.   I. International Atomic Energy
Agency.   II. Great Britain. Foreign and Commonwealth
Office.   III. Series: Foreign policy documents ; no.
136.
JX1974.73.R65   1986
327.1'74--dc19   86-228705

Sagan, Scott Douglas.
Sagan, Scott Douglas.
   The spread of nuclear weapons : a debate / Scott D.
Sagan, Kenneth N. Waltz. -- 1st ed. -- New York : W.W.
Norton, c1995.
   x, 160 p. : ill. ; 22 cm.
   Includes bibliographical references (p. [137]-154)
and index.
   ISBN 0-393-03810-6. -- ISBN 0-393-96716-6 (pbk.)
   1. Nuclear weapons.   2. Arms race.   3. Nuclear
nonproliferation.   I. Waltz, Kenneth Neal, 1924-
II. Title.
U264.S233   1995
355.02'17--dc20   94-24470

Sauer, Tom.
Sauer, Tom.
Nuclear arms control : nuclear deterrence in the post-cold war period / Tom Sauer. -- New York : St. Martin's Press, 1998.
p. cm.
Includes bibliographical references and index.
ISBN 0-312-21196-1 (cloth)
1. Nuclear arms control. 2. Nuclear nonproliferation. I. Title.
KZ5665.S28 1998
341.7'34--dc21   97-38222
CIP 9/98

Scheinman, Lawrence.
Scheinman, Lawrence.
The International Atomic Energy Agency and world nuclear order / Lawrence Scheinman. -- Washington, D.C. : Resources for the Future ; Baltimore : Distributed by the Johns Hopkins University Press, c1987.
xvi, 320 p. : ill. ; 23 cm.
Includes bibliographies and index.
ISBN 0-915707-35-7 (soft : alk. paper). -- ISBN 0-915707-36-5 (hard : alk. paper)
1. International Atomic Energy Agency. 2. Nuclear energy--Government policy--United States. 3. Nuclear nonproliferation. I. Title.
QC770.I4962S34 1987
333.79'26'0601--dc19   87-42832
r88

Scheinman, Lawrence.
Scheinman, Lawrence.
The nonproliferation role of the International Atomic Energy Agency : a critical assessment / Lawrence Scheinman. -- Washington, D.C. : Resources for the Future ; [Baltimore, Md.] : Distributed worldwide by John Hopkins University Press, 1985.
viii, 72 p. ; 23 cm.
Includes bibliographies.
ISBN 0-915707-18-7 (pbk.) : $8.95
1. Nuclear nonproliferation. 2. International Atomic Energy Agency. I. Resources for the Future. II. Title.
JX1974.73.S33 1985
327.1'74--dc19   85-42950

Schneider, Barry R.
Schneider, Barry R.
Future war and counterproliferation : U.S. military responses to NBC proliferation threats / Barry R. Schneider. -- Westport, CT : Praeger, 1999.
p. cm.
Includes bibliographical references and index.

Schneider, Barry R. (Cont'd)
Schneider, Barry R. (Cont'd)
ISBN 0-275-96278-4 (alk. paper)
1. Weapons of mass destruction. 2. Nuclear nonproliferation. 3. National security--United States. 4. United States--Military policy. I. Title.
U793.S36 1999
327.1'747'0973--dc21   98-26518
                                         CIP 1/99

Schneider, Barry R.
Schneider, Barry R.
Radical responses to radical regimes : evaluating preemptive counter-proliferation / Barry R. Schneider. -- Washington, D.C. : Institute for National Strategic Studies, National Defense University, 1995.
vi, 55 p. ; 23 cm. -- (McNair paper, ISSN 1071-7552 ; 41)
"May 1995."
Includes bibliographical references (p. 41-55).
1. Nuclear nonproliferation. 2. Disarmament. 3. Chemical weapons. 4. Biological weapons. 5. United States--Military policy. I. National Defense University. Institute for National Strategic Studies. II. Title. III. Series: McNair papers ; no. 41.
U411.I6S36 1995   95-187939

Schweitzer, Glenn E., 1930-
Schweitzer, Glenn E., 1930-
Moscow DMZ : the story of the international effort to convert Russian weapons science to peaceful purposes / Glenn E. Schweitzer. -- Armonk, N.Y. : M. E. Sharpe, c1996.
ix, 291 p. : ill. ; 24 cm.
Includes bibliographical references and index.
ISBN 1-56324-625-2 (alk. paper). -- ISBN 1-56324-626-0 (pbk. : alk. paper)
1. Economic conversion--Russia (Federation) 2. International Science and Technology Center. 3. Brain drain--Russia (Federation) 4. Nuclear nonproliferation. 5. United States--Relations--Russia (Federation) 6. Russia (Federation)--Relations--United States. I. Title.
HC340.12.Z9D446 1996
338.4'76233'0947--dc20   95-42353

Seaborg, Glenn Theodore, 1912-
Seaborg, Glenn Theodore, 1912-
Stemming the tide : arms control in the Johnson years / Glenn T. Seaborg with Benjamin S. Loeb. -- Lexington, Mass. : Lexington Books, c1987.
xxi, 495 p., [16] p. of plates : ill. ; 23 cm.
Bibliography: p. [473]-478.
Includes index.
ISBN 0-669-13105-9 (alk. paper)

Seaborg, Glenn Theodore, 1912- (Cont'd)
Seaborg, Glenn Theodore, 1912- (Cont'd)
1. Nuclear arms control--United States--History.
2. Nuclear arms control--Soviet Union--History.
3. Nuclear nonproliferation--History. 4. Johnson, Lyndon B. (Lyndon Baines), 1908-1973. I. Loeb, Benjamin S., 1914- . II. Title.
JX1974.7.S417 1987
327.1'74--dc19   86-40444

Security in Europe and North-East Asia Pacific :
Security in Europe and North-East Asia Pacific : cross-perspectives : first working conference, November 7-8, 1991, at IFRI, Paris / Institute français des relations internationales, the Japan Institute of International Affairs, Japanisch-Deutsches Zentrum Berlin. -- Tokyo, Japan : Japan Institute of International Affairs, [1992]
87 p. ; 30 cm. -- (JIIA paper, ISSN 0918-8843 ; 1992/no. 1)
Contributions presented at the first meeting between IFRI and JIIA.
Includes bibliographical references.
1. Nuclear nonproliferation. 2. Arms control--Europe. 3. Arms control--East Asia. 4. Security, International. I. Institute français des relations internationales. II. Nihon Kokusai Mondai Kenkyūjo. III. Japanisch-Deutsches Zentrum (Berlin, Germany) IV. Series.
JX1974.73.S43 1992
327.1'74--dc20   94-149047

Security in the new world order :
Security in the new world order : an Indo-French dialogue / edited by Dipankar Banerjee. -- New Delhi : Institute for Defence Studies and Analyses, 1994.
xv, 260 p. ; 23 cm.
I-E-96-903169; 63-92
Summary: Papers presented at a colloquium between Institute for Defence Studies and Analyses and Centre d'études et de recherches internationales, Fondation nationale des sciences politiques, at New Delhi, Dec. 1994.
Includes bibliographical references.
ISBN 8186019049
1. Securities, International--Congresses.
2. National security--South Asia--Congresses.
3. Nuclear nonproliferation--Congresses. 4. Post-communism--Communist countries--Congresses.
5. India--Foreign relations--France--Congresses.
6. France--Foreign relations--India--Congresses.
7. World politics--1989- --Congresses. I. Banerjee, Dipankar. II. Institute for Defence Studies and Analyses. III. Fondation nationale des sciences politiques. Centre d'études et de recherches internationales.

Security in the new world order : (Cont'd)
Security in the new world order : an ... (Cont'd)
JX1974.S429   1994   96-903169

Shaping nuclear policy for the 1990s :
Shaping nuclear policy for the 1990s : a compendium of
views : report of the Defense Policy Panel of the
Committee on Armed Services, House of Representatives,
One Hundred Second Congress, second session. --
Washington : U.S. G.P.O. : For sale by the U.S.
G.P.O., Supt. of Docs., Congressional Sales Office,
1993.
   viii, 632 p. : ill. ; 28 cm.
   At head of title: 102d Congress, 2d session.
Committee print. No. 14.
   Distributed to some depository libraries in
microfiche.
   Shipping list no.: 93-0233-P.
   "December 17, 1992."
   Supt. of Docs. no.: Y 4.AR 5/2:N 88/5
   GPO: 1012-A
   GPO: 1012-B (MF)
   ISBN 0-16-040067-8
   1. Nuclear nonproliferation. 2. Security,
International. 3. United States--Military policy.
4. World politics--1989-   I. United States.
Congress. House. Committee on Armed Services.
Defense Policy Panel.
UA23.S458   1993
327.1'74--dc20
DGPO/DLC
for Library of Congress   94-124676

Shyam Babu, D.
Shyam Babu, D.
   Nuclear non-proliferation : towards a universal NPT
regime / D. Shyam Babu. -- Delhi : Konark Publishers,
c1992.
   xviii, 206 p. ; 22 cm.
   I-E-69518
   Includes bibliographical references (p. [195]-201).
Includes index.
   ISBN 8122002927 : Rs200.00
   1. Nuclear nonproliferation. 2. Security,
International.   I. Title. II. Title: Nuclear
nonproliferation.
JX1974.73.S56   1992
327.1'74--dc20   92-907787

Sigal, Leon V.
Sigal, Leon V.
   Disarming strangers : nuclear diplomacy with North
Korea / Leon V. Sigal. -- Princeton, N.J. : Princeton
University Press, c1998.
   xi, 321 p. ; 25 cm. -- (Princeton studies in
international history and politics)
   Includes bibliographical references (p. [265]-305)

Sigal, Leon V. (Cont'd)
Sigal, Leon V. (Cont'd)
and index.
ISBN 0-691-05797-4 (cl : alk. paper)
1. Nuclear nonproliferation. 2. United States-
-Foreign relations--Korea (North) 3. Korea (North)-
-Foreign relations--United States. 4. Diplomacy.
I. Title. II. Series.
JZ5675.S55 1998
327.1'747--dc21 97-24502

Solomon, Kenneth A., 1947-
Solomon, Kenneth A., 1947-
Plutonium for Japan's nuclear reactors : paying both
the proliferation and dollar price to assure long-term
fuel supply / Kenneth Alvin Solomon. -- Santa Monica,
CA : Rand Corp., 1993.
xi, 32 p. : ill. ; 28 cm.
"MR-186-CC"--P. 4 of cover.
"Supported by the Carnegie Corporation."
Includes bibliographical references.
ISBN 0-8330-1367-X
1. Plutonium industry--Japan. 2. Nuclear industry-
-Japan. 3. Nuclear power plants--Japan. 4. Nuclear
nonproliferation. I. Carnegie Corporation of New
York. II. Title.
HD9539.P583J37 1993
338.4'762148335--dc20 93-10298

South Africa's policy on the non-proliferation of weapons of mass destruction :
South Africa's policy on the non-proliferation of
weapons of mass destruction : the role of the
Department of Foreign Affairs. -- Cape Town : South
African Communication Service on behalf of the Dept.
of Foreign Affais, 1995.
47 p. ; 21 cm.
ISBN 0-621-16974-9
1. Nuclear nonproliferation. 2. South Africa-
-Military policy. 3. South Africa--Foreign policy.
I. South Africa. Dept. of Foreign Affairs.
JZ5675.S68 1995
327.1'74'0968--dc21 97-176511

Spector, Leonard S.
Spector, Leonard S.
Going nuclear / Leonard S. Spector. -- Cambridge,
Mass. : Ballinger Pub. Co., c1987.
xv, 370 p., [4] p. of plates : ill., maps ; 21 cm.
"A Carnegie Endowment book."
Includes bibliographies and index.
ISBN 0-88730-145-2 (pbk.). -- ISBN 0-88730-144-4
1. Nuclear weapons. 2. Nuclear nonproliferation.
I. Title.
U264.S626 1987
355.8'25119--dc19 86-32115

Spector, Leonard S.
  Spector, Leonard S.
    Nuclear ambitions : the spread of nuclear weapons,
  1989-1990 / Leonard S. Spector with Jacqueline R.
  Smith. -- Boulder : Westview Press, 1990.
    450 p. : maps ; 24 cm.
    "A Carnegie endowment book."
    Includes bibliographical references (p. 305-416) and
  index.
    ISBN 0-8133-8075-8 (alk. paper). -- ISBN
  0-8133-8074-X
    1. Nuclear weapons. 2. World politics--1985-1995.
  3. Nuclear nonproliferation. I. Smith, Jacqueline R.
  II. Title.
  U264.S629   1990
  355.8'25119--dc20   90-12835

Spector, Leonard S.
  Spector, Leonard S.
    The undeclared bomb / Leonard S. Spector. --
  Cambridge, Mass. : Ballinger Pub. Co., c1988.
    xix, 499 p. : ill. ; 21 cm.
    "A Carnegie Endowment book."
    Includes bibliographical references.
    ISBN 0-88730-303-X. -- ISBN 0-88730-304-8 (pbk.)
    1. Nuclear weapons. 2. World politics--1985-1995.
  3. Nuclear nonproliferation. I. Title.
  U264.S636   1988
  355.8'25119--dc19   88-28726

Steinberg, Gerald M.
  Steinberg, Gerald M.
    Non-proliferation : time for regional approaches? /
  Gerald M. Steinberg. -- Ramat Gan, Israel : BESA
  Center, Bar-Ilan University, [1994]
    p. 419-423 ; 24 cm. -- (Security and policy studies,
  ISSN 0793-1042 ; no. 18)
    Reprinted from Orbis, v. 38, no. 3, Summer 1994.
    "September 1994."
    Includes bibliographical references.
    1. Nuclear nonproliferation. 2. Regionalism
  (International organization) I. Merkaz Besa le
  -meḥkarim asṭraṭegiyim. II. Title. III. Series.
  KZ5675.S8   1994
  327.1'747--dc21   97-215950
                                                                       r98

Subramanian, R. R. (Ram Rajan)
  Subramanian, R. R. (Ram Rajan)
    Nuclear competition in South Asia and U.S. policy /
  Ram R. Subramanian. -- Berkeley : Institute of
  International Studies, University of California,
  c1987.
    62 p. ; 24 cm. -- (Policy papers in international
  affairs ; no. 30)

Subramanian, R. R. (Ram Rajan) (Cont'd)
Subramanian, R. R. (Ram Rajan) (Cont'd)
Bibliography: p. 61-62.
ISBN 0-87725-530-X : $5.50
1. Nuclear nonproliferation. 2. Nuclear weapons-
-Pakistan. 3. United States--Foreign relations--South
Asia. 4. South Asia--Foreign relations--United
States. I. Title. II. Series.
JX1974.73.S83 1987
327.1'74--dc19 87-80669
r92

A Survey of European nuclear policy, 1985-87 /
A Survey of European nuclear policy, 1985-87 / edited by
Harald Müller. -- New York : St. Martin's Press, 1989.
xii, 158 p. : ill. ; 23 cm.
Includes bibliographies and index.
ISBN 0-312-02796-6 : $49.95 (est.)
1. Nuclear nonproliferation--History--Congresses.
2. Nuclear arms control--Europe--History--Congresses.
I. Müller, Harald, 1949 May 13-
JX1974.73.S87 1989
327.1'74--dc19 88-35933
r90

Symposium on International Safeguards (1994 : Vienna,
Austria)
Symposium on International Safeguards (1994 : Vienna,
Austria)
International nuclear safeguards, 1994--vision for
the future : proceedings of a Symposium on
International Safeguards organized in co-operation
with the American Nuclear Society ... [et al.], and
held in Vienna, 14-18 March 1994. -- Vienna :
International Atomic Energy Agency, 1994-
v. <1 > : ill. ; 24 cm. -- (Proceedings series,
ISSN 0074-1884)
"STI/PUB/945."
Includes bibliographical references.
ISBN 9201019947 (v. 1)
1. Nuclear facilities--Safety measures--Congresses.
2. Nuclear facilities--Security measures--Congresses.
3. Nuclear nonproliferation--Congresses.
I. International Atomic Energy Agency. II. American
Nuclear Society. III. Series: Proceedings series
(International Atomic Energy Agency)
TK9152.S855 1994
621.48'35--dc20 95-131973

Szilard, Leo.
Szilard, Leo.
Toward a livable world : Leo Szilard and the crusade
for nuclear arms control / edited by Helen S. Hawkins,
G. Allen Greb, Gertrud Weiss Szilard. -- Cambridge,
Mass. : MIT Press, c1987.
lxxiv, 499 p. ; 24 cm. -- (Collected works of Leo
Szilard ; v. 3)

Szilard, Leo. (Cont'd)
Szilard, Leo. (Cont'd)
"Bibliography of nonscientific works of Leo Szilard": p. [485]-487.
Includes bibliographies and index.
ISBN 0-262-19260-8. -- ISBN (invalid) 0-262-08162-8
1. Nuclear disarmament--History. 2. Nuclear nonproliferation--History. 3. Szilard, Leo.
I. Hawkins, Helen S. II. Greb, G. Allen.
III. Szilard, Gertrud Weiss. IV. Title. V. Series: Szilard, Leo. Works. 1978 ; v. 3.
QC3.S97 vol. 3
[JX1974.7]
539.7'092'4 s--dc19
[327.1'74'09]   86-18518

Tanter, Raymond.
Tanter, Raymond.
Rogue regimes : terrorism and proliferation / Raymond Tanter. -- 1st ed. -- New York : St. Martin's Press, 1998.
xiv, 331 p. ; 22 cm.
Includes bibliographical references (p. [261]-322) and index.
ISBN 0-312-17300-8
1. Dictators--Biography. 2. Khomeini, Ruhollah. 3. Hāshimī Rafsanjānī, ʿAlī Akbar. 4. Hussein, Saddam, 1937- . 5. Qaddafi, Muammar. 6. Assad, Hafez, 1928- . 7. Castro, Fidel, 1927- . 8. Kim, Chong-gil, 1926- 9. State-sponsored terrorism.
10. Nuclear nonproliferation. 11. National security--United States. 12. United States--Foreign relations--20th century. I. Title.
D412.7.T26 1998
327.1'17--dc21   97-21494

Taylor, Terence.
Taylor, Terence.
Escaping the prison of the past : rethinking arms control and non-proliferation measures / Terence Taylor. -- Stanford, Calif. : Center for International Security and Arms Control, Stanford University, c1996.
49, [14] p. ; 28 cm.
"April 1996."
Includes bibliographical references (p. 39-47).
ISBN 0-935371-43-5
1. Arms control. 2. Nuclear nonproliferation.
I. Stanford University. Center for International Security and Arms Control. II. Title.
KZ5624.T39 1996   97-133040

Tiwari, H. D. (Hari Dutt)
Tiwari, H. D. (Hari Dutt)
India and the problem of nuclear proliferation / H.D. Tiwari. -- 1st ed. -- Delhi : R.K. Publishers ; New Delhi : Distributed by D.K. Publishers'

Tiwari, H. D. (Hari Dutt) (Cont'd)
Tiwari, H. D. (Hari Dutt) (Cont'd)
Distributor, 1988.
xviii, 192 p. ; 23 cm.
I E 58998
Includes bibliographies and index.
Rs175.00
1. Nuclear nonproliferation. 2. Nuclear energy--India. I. Title.
JX1974.73.T58 1988
327.1'74'0954--dc20   89-900434

U.S. and Japanese nonproliferation export controls /
U.S. and Japanese nonproliferation export controls / edited by Gary K. Bertsch, Richard T. Cupitt, Takehiko Yamamoto. -- Lanham, MD : University Press of America, 1996.
p. cm.
Includes bibliographical references and index.
ISBN 0-7618-0190-1 (cloth : alk. paper). -- ISBN 0-7618-0191-X (paper : alk. paper)
1. Nuclear nonproliferation--International cooperation. 2. Export controls--United States. 3. Export controls--Japan. I. Bertsch, Gary K. II. Cupitt, Richard T. III. Yamamoto, Takehiko, 1943-
JX1974.73.U14 1996
327.1'74--dc20   95-43591
                                        CIP 1/96

U.S.-China commercial nuclear commerce :
U.S.-China commercial nuclear commerce : nonproliferation and trade issues / Project director: Robert E. Ebel. -- Washington, D.C. : Center for Strategic and International Studies, 1997.
ix, 34 p. : ill. ; 28 cm.
"September 1997."
Includes bibliographical references.
ISBN 0-89206-333-5
1. Nuclear industry--China. 2. Nuclear power plants--China. 3. Nuclear industry--United States. 4. Electric power plant equipment industry--United States. 5. Technical assistance, American--China. 6. China--Commerce--United States. 7. United States--Commerce--China. 8. Nuclear nonproliferation.
I. Ebel, Robert E. II. Center for Strategic and International Studies (Washington, D.C.)
HD9698.C62U18 1997   97-223550

**U.S.-Japan Study Group on Arms Control and Non
-Proliferation After the Cold War.**
U.S.-Japan Study Group on Arms Control and Non
-Proliferation After the Cold War.
Next steps in arms control and non-proliferation :
report of the U.S.-Japan Study Group on Arms Control
and Non-Proliferation After the Cold War / edited by
William Clark, Jr., and Ryukichi Imai ; co-sponsored
by Carnegie Endowment for International Peace [and]
International House of Japan. -- Washington, D.C. :
Carnegie Endowment for International Peace :
Distributed by Brookings Institution Press, c1996.
iv, 196 p. ; 23 cm.
Includes bibliographical references.
ISBN 0-87003-105-8
1. Nuclear nonproliferation. 2. Nuclear arms
control--East Asia. 3. Nuclear-weapon-free zones-
-East Asia. 4. Nuclear arms control--United States.
5. Nuclear weapons--Government policy--United States.
I. Clark, William, 1930- . II. Imai, Ryukichi,
1929- . III. Title.
JX1974.73.U16 1996
327.1'747--dc21   96-53440

**U.S.-Japan Study Group on Arms Control and Non
-Proliferation After the Cold War.**
U.S.-Japan Study Group on Arms Control and Non
-Proliferation After the Cold War.
The United States, Japan, and the future of nuclear
weapons : report of the U.S.-Japan Study Group on Arms
Control and Non-Proliferation After the Cold War. --
Washington, D.C. : Carnegie Endowment for
International Peace, c1995.
viii, 181 p. ; 23 cm.
"Co-sponsored by the Carnegie Endowment for
International Peace [and the] International House of
Japan."
"A Carnegie endowment book"--P. [4] of Cover.
Includes bibliographical references.
1. Nuclear nonproliferation. 2. Nuclear arms
control--United States. 3. Nuclear weapons-
-Government policy--Japan.
JZ5675.U16   1995
327.1'74--dc20   95-6459

U.S. nuclear non-proliferation policy, 1945-1991 :
U.S. nuclear non-proliferation policy, 1945-1991 : guide
and index / project director, Virginia I. Foran ;
series editors, Thomas S. Blanton ... [et al.]. --
[Washington, D.C.] : National Security Archive ;
Alexandria, Va. : Chadwyck-Healey, c1991.
2 v. : ill. ; 29 cm.
May serve as a guide to the microfiche collection
entitled: Nuclear non-proliferation, 1945-1991. 1992.
Microfiche 93/41 <MicRR>.

U.S. nuclear non-proliferation policy, ... (Cont'd)
U.S. nuclear non-proliferation policy, ... (Cont'd)
1. Nuclear nonproliferation--Indexes. 2. Nuclear disarmament--United States--Indexes. I. Blanton, Thomas S. II. National Security Archive (U.S.) III. Nuclear non-proliferation, 1945-1991. IV. Title: United States nuclear non-proliferation policy, 1945-1991.
JX1974.73.U18 1991
327.1'74'0973--dc20   92-240952
                                                           r93

The United Nations and nuclear non-proliferation /
The United Nations and nuclear non-proliferation / with an introduction by Boutros Boutros-Ghali. -- New York, NY : United Nations, Dept. of Public Information, c1995.
    199 p. : ill. ; 26 cm. -- (The United Nations blue books series ; v. 3)
    "United Nations publication sales No. E.95.I.17 (soft)"--T.p. verso.
    Includes indexes.
    ISBN 9211005574
    1. Nuclear nonproliferation. 2. United Nations. I. United Nations. II. Series.
JX1974.73.U53 1995
341.7'34--dc20
DLC/MoSU-L
    for Library of Congress   95-152663
                                                           r96

United States.
United States.
    Nuclear regulatory legislation. -- Washington, D.C. : U.S. Nuclear Regulatory Commission, Office of the Executive Legal Director : Office of the General Counsel : Available from GPO Sales Program, U.S. Nuclear Regulatory Commission ; Springfield, VA : National Technical Information Service [distributor], 1984-
    1 v. (loose-leaf) ; 23 cm.
    Compilation of nuclear legislation. Through 97th Congress, 2nd session.
    Distributed to depository libraries in microfiche.
    "June 1984."
    "Publication to be updated every two years (at the end of each Congress) by inserting or deleting of material"--Bibliographic data sheet.
    Includes bibliographical references.
    "NUREG-0980."
    Item 1051-H-2 (microfiche)
    Supt. of Docs. no.: Y 3.N 88:10/0980
    1. Nuclear energy--Law and legislation--United States. 2. Nuclear nonproliferation. I. U.S. Nuclear Regulatory Commission. Office of the Executive Legal Director. II. U.S. Nuclear Regulatory Commission. Office of the General Council.

United States. (Cont'd)
  United States. (Cont'd)
    III. Title.
    KF2138.A3 1984
    346.7304'67924--dc19
    [347.306467924]
    DGPO/DLC
    for Library of Congress   84-602951

United States. Congress. House. Committee on Energy and
Commerce. Special Subcommittee on U.S. Pacific Rim
Trade.
  United States. Congress. House. Committee on Energy
  and Commerce. Special Subcommittee on U.S. Pacific
  Rim Trade.
    Nuclear energy cooperation with China : hearing
  before the Speical Subcommittee on U.S.-Pacific Rim
  Trade of the Committee on Energy and Commerce, House
  of Representatives, Ninety-ninth Congress, first
  session, on the role of other nuclear energy supplier
  nations in the Pacific Rim, September 12, 1985. --
  Washington : U.S. G.P.O., 1986.
    iii, 175 p. ; 24 cm.
    Distributed to some depository libraries in
  microfiche.
    Shipping list no.: 86-396-P.
    Includes bibliographical references.
    "Serial no. 99-59."
    Item 1019-A, 1019-B (microfiche)
    Supt. of Docs. no.: Y 4.En 2/3:99-59
    1. Nuclear energy--United States. 2. Nuclear
  energy--China. 3. Nuclear nonproliferation.
  I. Title.
    KF27.E5547 1985a
    327.1'11'0973--dc19
    DGPO/DLC
    for Library of Congress   86-601766

United States. Congress. House. Committee on Energy and
Commerce. Special Subcommittee on U.S. Trade with China.
  United States. Congress. House. Committee on Energy
  and Commerce. Special Subcommittee on U.S. Trade with
  China.
    Nuclear energy cooperation with China : hearing
  before the Special Subcommittee on U.S. Trade with
  China of the Committee on Energy and Commerce, House
  of Representatives, Ninety-eighth Congress, second
  session, May 16, 1984. -- Washington : U.S. G.P.O.,
  1984.
    iii, 219 p. : ill., 1 map ; 24 cm.
    Distributed to some depository libraries in
  microfiche.
    Bibliography: p. 156.
    "Serial no. 98-148."
    Item 1019-A, 1019-B (microfiche)
    Supt. of Docs. no.: Y 4.En 2/3:98-148
    1. Nuclear energy--United States. 2. Nuclear

United States. Congress. House. Committee on ... (Cont'd)
United States. Congress. House. ... (Cont'd)
energy--China.   3. Nuclear nonproliferation.
4. United States--Foreign relations--China.   5. China-
-Foreign relations--United States.   I. Title.
KF27.E558 1984a
327.1'74--dc19
DGPO/DLC
for Library of Congress   84-604094

United States. Congress. House. Committee on Energy and Commerce. Subcommittee on Energy Conservation and Power.
United States. Congress. House. Committee on Energy and Commerce. Subcommittee on Energy Conservation and Power.
  Nuclear exports : hearing before the Subcommittee on Energy Conservation and Power of the Committee on Energy and Commerce, House of Representatives, Ninety-ninth Congress, second session, on the effectiveness of Department of Energy controls over the export of nuclear-related technology, information, and services, May 15, 1986. -- Washington : U.S. G.P.O. : For sale by the Supt. of Docs., Congressional Sales Office, U.S. G.P.O., 1987.
  iii, 71 p. ; 24 cm.
  Distributed to some depository libraries in microfiche.
  Shipping list no.: 87-43-P.
  "Serial no. 99-140."
  Item 1019-A, 1019-B (microfiche)
  Supt. of Docs. no.: Y 4.En 2/3:99-140
  1. Nuclear nonproliferation.   2. Export controls-
-United States.   3. Technology transfer--Law and legislation--United States.   4. United States. Dept. of Energy.   I. Title.
KF27.E5526 1986m
327.1'74--dc19
DGPO/DLC
for Library of Congress   87-600729
                                                       r98

United States. Congress. House. Committee on Energy and Commerce. Subcommittee on Oversight and Investigations.
United States. Congress. House. Committee on Energy and Commerce. Subcommittee on Oversight and Investigations.
  Nuclear nonproliferation : hearing before the Subcommittee on Oversight and Investigations of the Committee on Energy and Commerce, House of Representatives, One Hundred Second Congress, first session, concerning failed efforts to curtail Iraq's nuclear weapons program, April 24, 1991. -- Washington : U.S. G.P.O. : For sale by the U.S. G.P.O., Supt. of Docs., Congressional Sales Office, 1992.
  iv, 752 p. : ill. ; 24 cm.
  Item 1019-A, 1019-B (MF)
  Distributed to some depository libraries in

United States. Congress. House. Committee on ... (Cont'd)
United States. Congress. House. ... (Cont'd)
  microfiche.
   Shipping list no.: 92-288-P.
   "Serial no. 102-95."
   Supt. of Docs. no.: Y 4.En 2/3:102-95
   ISBN 0-16-038338-2
   1. Nuclear nonproliferation. 2. Nuclear arms control--Iraq. 3. Nuclear weapons--Iraq. 4. Nuclear weapons information, American. 5. United States. Dept. of Energy. I. Title.
   KF27.E5546 1991g
   327.1'74--dc20
   DGPO/DLC
   for Library of Congress   92-192542

United States. Congress. House. Committee on Foreign Affairs.
United States. Congress. House. Committee on Foreign Affairs.
   Consideration of authorizations for fiscal years 1992-93 for the former Soviet Republics; peacekeeping activities; implementation of the Salvadoran Peace Accords; and the Nonproliferation and Disarmament Fund : markup before the Committee on Foreign Affairs and its Subcommittee on International Operations, House of Representatives, One Hundred Second Congress, second session, on H.R. 4547, H.R. 4548, H.R. 4549, and H. Res. 391, February 25, March 11, and June 10, 1992. -- Washington : U.S. G.P.O. : For sale by the U.S. G.P.O., Supt. of Docs., Congressional Sales Office, 1992.
   iii, 311 p. ; 24 cm.
   Item 1017-A, 1017-B (MF)
   Distributed to some depository libraries in microfiche.
   Shipping list no.: 92-0547-P.
   Supt. of Docs. no.: Y 4.F 76/1:Au 8/5
   ISBN 0-16-039063-X
   1. Economic assistance, American--Former Soviet republics. 2. United Nations--Armed Forces--Finance. 3. El Salvador--Politics and government--1979-1992. 4. Nuclear nonproliferation. I. United States. Congress. House. Committee on Foreign Affairs. Subcommittee on International Operations. II. Title.
   KF27.F6 1992d
   327.1'11'0973--dc20
   DGPO/DLC
   for Library of Congress   92-252833
                                                   r94

United States. Congress. House. Committee on Foreign Affairs.
United States. Congress. House. Committee on Foreign Affairs.
Proliferation and arms control : hearings before the Committee on Foreign Affairs and its Subcommittee on Arms Control, International Security and Science, House of Representatives, One Hundred First Congress, second session, May 17 and July 11, 1990. -- Washington : U.S. G.P.O. : For sale by the Supt. of Docs., Congressional Sales Office, U.S. G.P.O., 1991.
iv, 404 p. : ill., maps ; 24 cm.
Item 1017-A, 1017-B (MF)
Distributed to some depository libraries in microfiche.
Shipping list no.: 91-225-P.
Includes bibliographical references.
Supt. of Docs. no.: Y 4.F 76/1:P 94/4
1. Arms control. 2. Arms race. 3. Nuclear nonproliferation--International cooperation. 4. National security--United States. I. United States. Congress. House. Committee on Foreign Affairs. Subcommittee on Arms Control, International Security and Science. II. Title.
KF27.F6 1990t
327.1'74--dc20
DGPO/DLC
for Library of Congress   92-168101
                                           r93

United States. Congress. House. Committee on Foreign Affairs.
United States. Congress. House. Committee on Foreign Affairs.
Proposed nuclear cooperation agreement with the People's Republic of China : hearing and markup before the Committee on Foreign Affairs, House of Representatives, Ninety-ninth Congress, first session on H.J. Res. 404, July 31; November 13, 1985. -- Washington : U.S. G.P.O. : For sale by the Supt. of Docs., Congressional Sales Office, U.S. G.P.O., 1987.
iii, 243 p. ; 24 cm.
Distributed to some depository libraries in microfiche.
Shipping list no.: 87-504-P.
Includes bibliographical references.
Item 1017-A, 1017-B (microfiche)
Supt. of Docs. no.: Y 4.F 76/1:N 88/16
1. Nuclear energy--Law and legislation--United States. 2. Nuclear energy--Law and legislation--China. 3. Nuclear nonproliferation. I. Title.

United States. Congress. House. Committee on ... (Cont'd)
United States. Congress. House. ... (Cont'd)
KF27.F6   1985x
341.7'55--dc20
DGPO/DLC
for Library of Congress   87-602227

United States. Congress. House. Committee on Foreign
Affairs.
United States. Congress. House. Committee on Foreign
Affairs.
   The South Pacific nuclear free zone : hearings and
markup before the Committee on Foreign Affairs and its
Subcommittee on Asian and Pacific Affairs, House of
Representatives, One Hundredth Congress, first session
on, H. Con. Res. 158, June 9, and July 15, 1987. --
Washington : U.S. G.P.O. : For sale by the Supt. of
Docs., Congressional Sales Office, U.S. G.P.O., 1988.
     iii, 140 p. : 1 map ; 24 cm.
   Distributed to some depository libraries in
microfiche.
     Shipping list no.: 88-482-P.
     Item 1017-A, 1017-B (microfiche)
     Supt. of Docs. no.: Y 4.F 76/1:N 88/17
   1. Nuclear nonproliferation. 2. Nuclear-weapon-free
zones--Oceania. 3. United States--Foreign relations-
-Oceania. 4. Oceania--Foreign relations--United
States. I. United States. Congress. House.
Committee on Foreign Affairs. Subcommittee on Asian
and Pacific Affairs. II. Title.
KF27.F6   1987p
327.1'74'099--dc19
DGPO/DLC
for Library of Congress   88-602392
                                            r90

United States. Congress. House. Committee on Foreign
Affairs.
United States. Congress. House. Committee on Foreign
Affairs.
   U.S. nonproliferation policy : hearing before the
Committee on Foreign Affairs, House of
Representatives, One Hundred Third Congress, first
session, November 10, 1993. -- Washington : U.S.
G.P.O. : For sale by the U.S. G.P.O., Supt. of Docs.,
Congressional Sales Office, 1994.
     iii, 89 p. ; 24 cm.
   Distributed to some depository libraries in
microfiche.
     Shipping list no.: 94-0083-P.
     Supt. of Docs. no.: Y 4.F 76/1:N 73/4
     GPO: 1017-A
     GPO: 1017-B (MF)
     ISBN 0-16-043685-0
   1. Nuclear nonproliferation. 2. Export controls-
-United States. 3. Nuclear arms control--Government
policy--United States. I. Title. II. Title: US

United States. Congress. House. Committee on ... (Cont'd)
United States. Congress. House. ... (Cont'd)
nonproliferation policy.
KF27.F6 1993m
DGPO/DLC
for Library of Congress 94-174860

United States. Congress. House. Committee on Foreign Affairs.
United States. Congress. House. Committee on Foreign Affairs.
U.S. nuclear policy : hearing before the Committee on Foreign Affairs, House of Representatives, One Hundred Third Congress, second session, October 5, 1994. -- Washington : U.S. G.P.O. : For sale by the U.S. G.P.O., Supt. of Docs., Congressional Sales Office, 1995.
ii, 97 p. : ill. ; 24 cm.
Distributed to some depository libraries in microfiche.
Shipping list no.: 95-0098-P.
Supt. of Docs. no.: Y 4.F 76/1:N 88/25
GPO: 1017-A
GPO: 1017-B (MF)
ISBN 0-16-046851-5
1. Nuclear weapons--Government policy--United States. 2. United States--Military policy.
3. Nuclear arms control--United States. 4. Nuclear nonproliferation. I. Title.
KF27.F6 1994o 95-184806

United States. Congress. House. Committee on Foreign Affairs.
United States. Congress. House. Committee on Foreign Affairs.
United States-Japan Nuclear Cooperation Agreement : hearings before the Committee on Foreign Affairs, House of Representatives, One Hundredth Congress, first and second sessions, December 16, 1987 and March 2, 1988. -- Washington : U.S. G.P.O. : For sale by the Supt. of Docs., Congressional Sales Office, U.S. G.P.O., 1988.
v, 706 p. : ill. ; 24 cm.
Distributed to some depository libraries in microfiche.
Shipping list no.: 88-484-P.
Bibliography: p. 382.
Item 1017-A, 1017-B (microfiche)
Supt. of Docs. no.: Y 4.F 76/1:N 88/18
1. Nuclear nonproliferation. 2. Nuclear energy--Law and legislation--United States. 3. Nuclear energy--Law and legislation--Japan. I. Title. II. Title: United States Japan Nuclear Cooperation Agreement.

United States. Congress. House. Committee on ... (Cont'd)
United States. Congress. House. ... (Cont'd)
KF27.F6 1988
346.7304'67924--dc19
[347.306467924]
DGPO/DLC
for Library of Congress   88-602468

United States. Congress. House. Committee on Foreign
Affairs. Subcommittee on Arms Control, International
Security, and Science.
United States. Congress. House. Committee on Foreign
Affairs. Subcommittee on Arms Control, International
Security, and Science.
   Missile proliferation : the need for controls
(Missile Technology Control Regime) : hearing before
the Subcommittees on Arms Control, International
Security and Science, and on International Economic
Policy and Trade of the Committee on Foreign Affairs,
House of Representatives, One Hundred First Congress,
first session, July 12, October 30, 1989. --
Washington : U.S. G.P.O. : For sale by the Supt. of
Docs., Congressional Sales Office, U.S. G.P.O., 1990.
   iii, 217 p. : ill., map ; 24 cm.
   Distributed to some depository libraries in
microfiche.
   Shipping list no.: 90-324-P.
   Includes bibliographical references.
   Item 1017-A, 1017-B (MF)
   Supt. of Docs. no.: Y 4.F 76/1:M 69/4
   1. Ballistic missiles--Government policy--United
States.   2. Nuclear nonproliferation.   3. Nuclear arms
control.   I. United States. Congress. House.
Committee on Foreign Affairs. Subcommittee on
International Economic Policy and Trade.   II. Title.
KF27.F636 1989b
DGPO/DLC
for Library of Congress   90-601260

United States. Congress. House. Committee on Foreign
Affairs. Subcommittee on Arms Control, International
Security, and Science.
United States. Congress. House. Committee on Foreign
Affairs. Subcommittee on Arms Control, International
Security, and Science.
   North Korean nuclear program : joint briefing before
the Subcommittees on Arms Control, International
Security and Science; Asian and Pacific Affairs; and
International Economic Policy and Trade of the
Committee on Foreign Affairs, House of
Representatives, One Hundred Second Congress, second
session, July 22, 1992. -- Washington : U.S. G.P.O. :
For sale by the U.S. G.P.O., Supt. of Docs.,
Congressional Sales Office, 1992.
   v, 34 p. ; 23 cm.
   ISBN 0-16-039758-8
   1. Nuclear nonproliferation.   2. Nuclear arms

United States. Congress. House. Committee on ... (Cont'd)
United States. Congress. House. ... (Cont'd)
control--Korea (North)--Verification. 3. United
States--Foreign relations--Korea (North) 4. Korea
(North)--Foreign relations--United States. 5. Nuclear
nonproliferation. I. United States. Congress.
House. Committee on Foreign Affairs. Subcommittee on
Asian and Pacific Affairs. II. United States.
Congress. House. Committee on Foreign Affairs.
Subcommittee on International Economic Policy and
Trade. III. Title.
KF27.F636 1992a
341.7'34'095193--dc20
DGPO
for Library of Congress   93-125592

United States. Congress. House. Committee on Foreign Affairs. Subcommittee on Arms Control, International Security, and Science.
United States. Congress. House. Committee on Foreign
Affairs. Subcommittee on Arms Control, International
Security, and Science.
Pakistan and United States nuclear nonproliferation
policy : hearing before the Subcommittee on Arms
Control, International Security, and Science, and
Asian and Pacific Affairs, and International Economic
Policy and Trade of the Committee on Foreign Affairs,
House of Representatives, One Hundredth Congress,
first session, October 22, 1987. -- Washington : U.S.
G.P.O. : For sale by the Supt. of Docs., Congressional
Sales Office, U.S. G.P.O., 1988.
v, 172 p. : ill., maps ; 24 cm.
Distributed to some depository libraries in
microfiche.
Shipping list no.: 88-566-P.
Bibliography: p. 168-172.
Item 1017-A, 1017-B (microfiche)
Supt. of Docs. no.: Y 4.F 76/1:P 17/9
1. Nuclear nonproliferation. 2. Pakistan--Foreign
relations--United States. 3. United States--Foreign
relations--Pakistan. I. United States. Congress.
House. Committee on Foreign Affairs. Subcommittee on
Asian and Pacific Affairs. II. United States.
Congress. House. Committee on Foreign Affairs.
Subcommittee on International Economic Policy and
Trade. III. Title.
KF27.F636 1987e
327.1'74--dc19
DGPO/DLC
for Library of Congress   88-602805

United States. Congress. House. Committee on Foreign
Affairs. Subcommittee on Arms Control, International
Security, and Science.
United States. Congress. House. Committee on Foreign
Affairs. Subcommittee on Arms Control, International
Security, and Science.
Proliferation and arms control in the 1990's :
hearing before the Subcommittee on Arms Control,
International Security, and Science of the Committee
on Foreign Affairs, House of Representatives, One
Hundred Second Congress, second session, March 3,
1992. -- Washington : U.S. G.P.O. : For sale by the
U.S. G.P.O., Supt. of Docs., Congressional Sales
Office, 1993.
iii, 85 p. ; 24 cm.
Distributed to some depository libraries in
microfiche.
Shipping list no.: 93-0219-P.
Supt. of Docs. no.: Y 4.F 76/1:P 94/5
GPO: 1017-A
GPO: 1017-B (MF)
ISBN 0-16-039963-7
1. Arms control. 2. Arms race. 3. Nuclear
nonproliferation--International cooperation.
4. National security--United States. I. Title.
KF27.F636 1992c
327.1'74--dc20
DGPO/DLC
for Library of Congress   93-198484

United States. Congress. House. Committee on Foreign
Affairs. Subcommittee on Arms Control, International
Security, and Science.
United States. Congress. House. Committee on Foreign
Affairs. Subcommittee on Arms Control, International
Security, and Science.
Third Review Conference of the Treaty on the Non
-Proliferation of Nuclear Weapons : hearing before the
subcommittees on Arms Control, International Security,
and Science and on International Economic Policy and
Trade of the Committee on Foreign Affairs, House of
Representatives, Ninety-ninth Congress, first session,
August 1, 1985. -- Washington : U.S. G.P.O. : For sale
by the Supt. of Docs., Congressional Sales Office,
U.S. G.P.O., 1986.
iii, 165 p.: ill. ; 24 cm.
Distributed to some depository libraries in
microfiche.
Shipping list no.: 86-514-P.
Includes bibliographical references.
Item 1017-A, 1017-B (microfiche)
Supt. of Docs. no.: Y 4.F 76/1:N 88/3/985
1. Nuclear nonproliferation--Congresses. 2. Non
-Proliferation Treaty Review Conference (3rd : 1985 :
Geneva, Switzerland)   I. United States. Congress.

United States. Congress. House. Committee on ... (Cont'd)
United States. Congress. House. ... (Cont'd)
House. Committee on Foreign Affairs. Subcommittee on International Economic Policy and Trade. II. Title.
KF27.F636   1985b
341.7'54--dc19
DGPO/DLC
for Library of Congress   86-601915

United States. Congress. House. Committee on Foreign Affairs. Subcommittee on Asia and the Pacific.
United States. Congress. House. Committee on Foreign Affairs. Subcommittee on Asia and the Pacific.
   Concerning the establishment of a South Pacific nuclear free zone and expressing the sense of the Congress with respect to the South Pacific region : markups before the Subcommittee on Asia and the Pacific of the Committee on Foreign Affairs, House of Representatives, One Hundred Third Congress, first session, on H. Con. Res. 111 and H. Con. Res. 180, November 16, 1993. -- Washington : U.S. G.P.O. : For sale by the U.S. G.P.O., Supt. of Docs., Congressional Sales Office, 1995.
   iii, 17 p. ; 24 cm.
   Distributed to some depository libraries in microfiche.
     Shipping list no.: 95-0085-P.
     Supt. of Docs. no.: Y 4.F 76/1:N 88/24
     GPO: 1017-A
     GPO: 1017-B (MF)
     ISBN 0-16-046846-9
   1. Nuclear-weapon-free zones--Oceania. 2. Nuclear nonproliferation. 3. United States--Foreign relations--Oceania. 4. Oceania--Foreign relations--United States.   I. Title.
KF27.F638   1993d
DGPO/DLC
for Library of Congress   95-178341

United States. Congress. House. Committee on Foreign Affairs. Subcommittee on Asia and the Pacific.
United States. Congress. House. Committee on Foreign Affairs. Subcommittee on Asia and the Pacific.
   Developments in North Korea : hearing before the Subcommittee on Asia and the Pacific of the Committee on Foreign Affairs, House of Representatives, One Hundred Third Congress, second session, June 9, 1994. -- Washington : U.S. G.P.O. : For sale by the U.S. G.P.O., Supt. of Docs., Congressional Sales Office, 1995.
   iii, 32 p. ; 24 cm.
   Distributed to some depository libraries in microfiche.
     Shipping list no.: 95-0098-P.
     Supt. of Docs. no.: Y 4.F 76/1:K 84/16
     GPO: 1017-A
     GPO: 1017-B (MF)

United States. Congress. House. Committee on ... (Cont'd)
United States. Congress. House. ... (Cont'd)
   ISBN 0-16-046844-2
   1. Nuclear nonproliferation. 2. Nuclear weapons-
-Korea (North)   3. Nuclear arms control--Verification-
-Korea (North)   I. Title.
   KF27.F638   1994d
   DGPO/DLC
   for Library of Congress   95-171939

United States. Congress. House. Committee on Foreign
Affairs. Subcommittee on Asian and Pacific Affairs.
United States. Congress. House. Committee on Foreign
Affairs. Subcommittee on Asian and Pacific Affairs.
   The implications of the Arshad Pervez case for U.S.
policy toward Pakistan : hearing before the
Subcommittee on Asian and Pacific Affairs and on
International Economic Policy and Trade of the
Committee on Foreign Affairs, House of
Representatives, One Hundredth Congress, second
session, February 17, 1988. -- Washington : U.S.
G.P.O. : For sale by the Supt. of Docs., Congressional
Sales Office, U.S. G.P.O., 1989.
   iii, 83 p. ; 23 cm
   Distributed to some depository libraries in
microfiche.
   Shipping list no.: 89-323-P.
   Item 1017-A, 1017-B (microfiche)
   Supt. of Docs. no.: Y 4.F 76/1:P 43/10
   1. Nuclear nonproliferation. 2. Pervez, Arshad.
   3. United States--Foreign relations--Pakistan.
   4. Pakistan--Foreign relations--United States.
   I. United States. Congress. House. Committee on
Foreign Affairs. Sucommittee on International
Economic Policy and Trade. II. Title.
   KF27.F638   1988k
   327.7305491--dc20
   DGPO/DLC
   for Library of Congress   89-602130

United States. Congress. House. Committee on Foreign
Affairs. Subcommittee on Asian and Pacific Affairs.
United States. Congress. House. Committee on Foreign
Affairs. Subcommittee on Asian and Pacific Affairs.
   Korea : North-South nuclear issues : hearing before
the Subcommittee on Asian and Pacific Affairs of the
Committee on Foreign Affairs, House of
Representatives, One Hundred First Congress, second
session, July 25, 1990. -- Washington : U.S. G.P.O. :
For sale by the Supt. of Docs., Congressional Sales
Office, U.S. G.P.O., 1991.
   iii, 93 p. ; 24 cm.
   Includes bibliographical references.
   1. Nuclear weapons--Korea (North) 2. Nuclear
nonproliferation. 3. United States--Foreign
relations--Korea (North) 4. Korea (North)--Foreign
relations--United States. I. Title.

United States. Congress. House. Committee on ... (Cont'd)
United States. Congress. House. ... (Cont'd)
KF27.F638   1990n   91-600861

United States. Congress. House. Committee on Foreign
Affairs. Subcommittee on Asian and Pacific Affairs.
United States. Congress. House. Committee on Foreign
Affairs. Subcommittee on Asian and Pacific Affairs.
   Pakistan's illegal nuclear procurement in the United
States : hearing before the subcommittees on Asian and
Pacific Affairs and on International Economic Policy
and Trade of the Committee on Foreign Affairs, House
of Representatives, One Hundredth Congress, first
session, July 22, 1987. -- Washington : U.S. G.P.O. :
For sale by the Supt. of Docs., Congressional Sales
Office, U.S. G.P.O., 1988.
   iii, 39 p. ; 24 cm.
   Distributed to some depository libraries in
microfiche.
   Shipping list no.: 88-602-P.
   Item 1017-A, 1017-B (microfiche)
   Supt. of Docs. no.: Y 4.F 76/1:P 17/10
   1. Nuclear nonproliferation. 2. Export controls-
-United States. 3. Nuclear weapons--Pakistan.
4. Economic sanctions, American--Pakistan. I. United
States. Congress. House. Committee on Foreign
Affairs. Subcommittee on International Economic
Policy and Trade. II. Title.
KF27.F638   1987f
327.1'74--dc19
DGPO/DLC
   for Library of Congress   88-602934

United States. Congress. House. Committee on Foreign
Affairs. Subcommittee on Asian and Pacific Affairs.
United States. Congress. House. Committee on Foreign
Affairs. Subcommittee on Asian and Pacific Affairs.
   Policy implications of North Korea's ongoing nuclear
program and markup of H. Con. Res. 179, H. Con. Res.
189, and H. Con. Res. 240 : hearing and markup before
the Subcommittee on Asian and Pacific Affairs of the
Committee on Foreign Affairs, House of
Representatives, One Hundred Second Congress, first
session, November 21, 1991. -- Washington : U.S.
G.P.O. : For sale by the U.S. G.P.O., Supt. of Docs.,
Congressional Sales Office, 1992.
   iii, 75 p. ; 24 cm.
   Item 1017-A, 1017-B (MF)
   Distributed to some depository libraries in
microfiche.
   Shipping list no.: 92-0473-P.
   Supt. of Docs. no.: Y 4.F 76/1:N 81/17
   ISBN 0-16-038829-5
   1. Nuclear weapons--Korea (North) 2. Nuclear
facilities--Korea (North) 3. Nuclear
nonproliferation. 4. United States--Foreign
relations--Korea (North) 5. Korea (North)--Foreign

United States. Congress. House. Committee on ... (Cont'd)
United States. Congress. House. ... (Cont'd)
relations--United States.   I. Title.
Kf27.F638   1991a
347.73'0412--dc20
[347.302412]
DGPO/DLC
for Library of Congress   92-222989

United States. Congress. House. Committee on Foreign
Affairs. Subcommittee on Asian and Pacific Affairs.
United States. Congress. House. Committee on Foreign
Affairs. Subcommittee on Asian and Pacific Affairs.
   Political developments in Pakistan : hearing before
the Subcommittee on Asian and Pacific Affairs of the
Committee on Foreign Affairs, House of
Representatives, One Hundred First Congress, second
session, October 2, 1990. -- Washington : U.S. G.P.O.
: For sale by the Supt. of Docs., Congressional Sales
Office, U.S. G.P.O., 1991.
   iii, 85 p. ; 24 cm.
   Distributed to some depository libraries in
microfiche.
   Shipping list no.: 91-397-P.
   Item 1017-A, 1017-B (MF)
   Includes bibliographical references.
   Supt. of Docs. no.: Y 4.F 76/1:P 75/25
   1. Pakistan--Politics and government--1988-
   2. United States--Foreign relations--Pakistan.
   3. Pakistan--Foreign relations--United States.
   4. Nuclear nonproliferation.   I. Title.
KF27.F638   19901
DGPO/DLC
for Library of Congress   91-601047

United States. Congress. House. Committee on Foreign
Affairs. Subcommittee on Europe and the Middle East.
United States. Congress. House. Committee on Foreign
Affairs. Subcommittee on Europe and the Middle East.
   Iraq's nuclear weapons capability and IAEA
inspections in Iraq : joint hearing before the
Subcommittees on Europe and the Middle East and
International Security, International Organizations,
and Human Rights of the Committee on Foreign Affairs,
House of Representatives, One Hundred Third Congress,
first session, June 29, 1993. -- Washington : U.S.
G.P.O. : For sale by the U.S. G.P.O., Supt. of Docs.,
Congressional Sales Office, 1993.
   iii, 206 p. : ill. ; 24 cm.
   Distributed to some depository libraries in
microfiche.
   Shipping list no.: 93-0660-P.
   Includes bibliographical references.
   Supt. of Docs. no.: Y 4.F 76/1:IR 1/13
   GPO: 1017-A
   GPO: 1017-B (MF)
   ISBN 0-16-041691-4

United States. Congress. House. Committee on ... (Cont'd)
United States. Congress. House. ... (Cont'd)
1. Nuclear weapons--Iraq. 2. Nuclear
nonproliferation. 3. Nuclear facilities--Iraq.
I. United States. Congress. House. Committee on
Foreign Affairs. Subcommittee on International
Security, International Organizations, and Human
Rights. II. Title.
KF27.F64214 1993a
327.730567'09'049--dc20
DGPO/DLC
for Library of Congress 94-115646

United States. Congress. House. Committee on Foreign Affairs. Subcommittee on International Security and Scientific Affairs.
United States. Congress. House. Committee on Foreign
Affairs. Subcommittee on International Security and
Scientific Affairs.
Proposed amendments to the Nuclear Non-proliferation
Act, 1983 : hearings before the Committee on Foreign
Affairs and its Subcommittees on International
Security and Scientific Affairs and on International
Economic Policy and Trade of the House of
Representatives, Ninety-eighth Congress, first
session, on H.R. 1417 and H.R. 3058, September 20;
October 20, 26; November 1, 1983. -- Washington : U.S.
G.P.O., 1984.
iv, 382 p. : ill. ; 24 cm.
Distributed to some depository libraries in
microfiche.
Item 1017-A, 1017-B (microfiche)
Supt. of Docs. no.: Y 4.F 76/1:N 88/11/983
1. Nuclear nonproliferation. I. United States.
Congress. House. Committee on Foreign Affairs.
Subcommittee on International Economic Policy and
Trade. II. Title.
KF27.F64825 1983b
346.7304'67924--dc19
[347.306467924]
DGPO/DLC
for Library of Congress 84-602897

United States. Congress. House. Committee on Foreign Affairs. Subcommittee on International Security, International Organizations, and Human Rights.
United States. Congress. House. Committee on Foreign
Affairs. Subcommittee on International Security,
International Organizations, and Human Rights.
The Arms Control and Disarmament Agency : hearing
before the Subcommittee on International Security,
International Organizations, and Human Rights of the
Committee on Foreign Affairs, House of
Representatives, One Hundred Third Congress, first
session, April 27, 1993. -- Washington : U.S. G.P.O. :
For sale by the U.S. G.P.O., Supt. of Docs.,

United States. Congress. House. Committee on ... (Cont'd)
United States. Congress. House. ... (Cont'd)
Congressional Sales Office, 1994.
   iii, 25 p. ; 24 cm.
   Distributed to some depository libraries in microfiche.
   Shipping list no.: 94-0107-P.
   Supt. of Docs. no.: Y 4.F 76/1:AR 5/47
   GPO: 1017-A
   GPO: 1017-B (MF)
   ISBN 0-16-043983-3
   1. United States. Arms Control and Disarmament Agency. 2. Nuclear nonproliferation. 3. Arms control--Verification. 4. Treaty on the Non-proliferation of Nuclear Weapons (1968)   I. Title.
   KF27.F645   1993f
   DGPO/DLC
   for Library of Congress   94-176138

United States. Congress. House. Committee on Foreign Affairs. Subcommittee on International Security, International Organizations, and Human Rights.
United States. Congress. House. Committee on Foreign Affairs. Subcommittee on International Security, International Organizations, and Human Rights.
   Challenges to U.S. security in the 1990's : hearing before the Subcommittee on International Security, International Organizations, and Human Rights of the Committee on Foreign Affairs, House of Representatives, One Hundred Third Congress, second session, March 17, April 21, June 9, June 27, and August 1, 1994. -- Washington : U.S. G.P.O. : For sale by the U.S. G.P.O., Supt. of Docs., Congressional Sales Office, 1994 [i.e. 1995]
   iii, 187 p. ; 24 cm.
   Distributed to some depository libraries in microfiche.
   Shipping list no.: 95-0075-P.
   Supt. of Docs. no.: Y 4.F 76/1:SE 2/24
   GPO: 1017-A
   GPO: 1017-B (MF)
   ISBN 0-16-046456-0
   1. National security--United States. 2. United States--Foreign relations. 3. Nuclear nonproliferation. 4. Nuclear weapons.   I. Title.
   KF27.F645   1994s
   DGPO/DLC
   for Library of Congress   95-184312

United States. Congress. House. Committee on Foreign Affairs. Subcommittee on International Security, International Organizations, and Human Rights.
United States. Congress. House. Committee on Foreign Affairs. Subcommittee on International Security, International Organizations, and Human Rights.
Concerning the establishment of a South Pacific nuclear free zone and concerning the emancipation of the Iranian Baha'i community : markups before the Subcommittee on International Security, International Organizations, and Human Rights of the Committee on Foreign Affairs, House of Representatives, One Hundred Third Congress, second session, on H. Con. Res. 111 and H. Con. Res. 124, February 3, 1994. -- Washington : U.S. G.P.O. : For sale by the U.S. G.P.O., Supt. of Docs., Congressional Sales Office, 1995.
iii, 18 p. ; 24 cm.
Distributed to some depository libraries in microfiche.
Shipping list no.: 95-0085-P.
Supt. of Docs. no.: Y 4.F 76/1:N 88/23
GPO: 1017-A
GPO: 1017-B (MF)
ISBN 0-16-046860-4
1. Nuclear-weapon-free zones--Oceania. 2. Nuclear nonproliferation. 3. Bahais--Civil rights--Iran. 4. Human rights--Iran. I. Title.
KF27.F645 1994o
DGPO/DLC
for Library of Congress   95-184308

United States. Congress. House. Committee on Foreign Affairs. Subcommittee on International Security, International Organizations, and Human Rights.
United States. Congress. House. Committee on Foreign Affairs. Subcommittee on International Security, International Organizations, and Human Rights.
A revitalized ACDA in the post-cold war world : joint hearing before the Subcommittee on International Security, International Organizations, and Human Rights and International Operations of the Committee on Foreign Affairs, House of Representatives, One Hundred Third Congress, second session, June 23, 1994. -- Washington : U.S. G.P.O. : For sale by the U.S. G.P.O., Supt. of Docs., Congressional Sales Office, 1994.
iii, 64 p. ; 24 cm.
Distributed to some depository libraries in microfiche.
Shipping list no.: 94-0381-P.
Supt. of Docs. no.: Y 4.F 76/1:P 84/10
GPO: 1017-A
GPO: 1017-B (MF)
ISBN 0-16-045911-7
1. Nuclear nonproliferation. 2. Treaty on the Non

United States. Congress. House. Committee on ... (Cont'd)
United States. Congress. House. ... (Cont'd)
-proliferation of Nuclear Weapons (1968) 3. United
States. Arms Control and Disarmament Agency.
4. National security--United States. I. United
States. Congress. House. Committee on Foreign
Affairs. Subcommittee on International Operations.
II. Title.
KF27.F645 1994i
327.1'74--dc20
DGPO/DLC
for Library of Congress   95-196821

United States. Congress. House. Committee on Foreign Affairs. Subcommittee on International Security, International Organizations, and Human Rights.
United States. Congress. House. Committee on Foreign
Affairs. Subcommittee on International Security,
International Organizations, and Human Rights.
  The security situation on the Korean Peninsula :
joint hearing before the Subcommittee on International
Security, International Organizations, and Human
Rights and Asia and the Pacific of the Committee on
Foreign Affairs, House of Representatives, One Hundred
Third Congress, second session, February 24, 1994. --
Washington : U.S. G.P.O. : For sale by the U.S.
G.P.O., Supt. of Docs., Congressional Sales Office,
1994.
  v, 80 p. : map ; 24 cm.
  Distributed to some depository libraries in
microfiche.
  Shipping list no.: 94-0234-P.
  Supt. of Docs. no.: Y 4.F 76/1:K 84/14
  GPO: 1017-A
  GPO: 1017-B (MF)
  ISBN 0-16-044491-8
  1. Korea (North)--Politics and government.
2. Nuclear nonproliferation. 3. Arms control-
-Verification. 4. Treaty on the Non-proliferation of
Nuclear Weapons (1968) I. United States. Congress.
House. Committee on Foreign Affairs. Subcommittee on
Asia and the Pacific. II. Title.
KF27.F645 1994a
DGPO/DLC
for Library of Congress   94-199332

United States. Congress. House. Committee on Foreign Affairs. Subcommittee on International Security, International Organizations, and Human Rights.
United States. Congress. House. Committee on Foreign Affairs. Subcommittee on International Security, International Organizations, and Human Rights.
    U.S. security policy toward rogue regimes : hearings before the Subcommittee on International Security, International Organizations, and Human Rights of the Committee on Foreign Affairs, House of Representatives, One Hundred Third Congress, first session, July 28 and September 14, 1993. -- Washington : U.S. G.P.O. : For sale by the U.S. G.P.O., Supt. of Docs., Congressional Sales Office, 1994.
    iii, 169 p. ; 24 cm.
    Distributed to some depository libraries in microfiche.
    Shipping list no.: 94-0107-P.
    Supt. of Docs. no.: Y 4.F 76/1:SE 2/23
    GPO: 1017-A
    GPO: 1017-B (MF)
    ISBN 0-16-043984-1
    1. Terrorism. 2. Nuclear warfare. 3. Nuclear nonproliferation. 4. Security, International. I. Title. II. Title: US security policy toward rogue regimes.
    KF27.F645 1993d
    DGPO/DLC
    for Library of Congress   94-182674

United States. Congress. House. Committee on International Relations.
United States. Congress. House. Committee on International Relations.
    H.R. 361, the Omnibus Export Administration Act of 1995 : markup before the Committee on International Relations, House of Representatives, One Hundred Fourth Congress, second session, March 29, 1996. -- Washington : U.S. G.P.O. : For sale by the U.S. G.P.O., Supt. of Docs., Congressional Sales Office, 1996.
    iii, 232 p. ; 24 cm.
    Distributed to some depository libraries in microfiche.
    Shipping list no.: 96-0357-P.
    Supt. of Docs. no.: Y 4.IN 8/16:OM 5
    GPO: 1017-A-01
    GPO: 1017-B-01 (MF)
    ISBN 0-16-052996-4
    1. Export controls--United States. 2. Technology transfer--Law and legislation--United States. 3. Nuclear nonproliferation. 4. Weapons of mass destruction. I. Title.

United States. Congress. House. Committee on ... (Cont'd)
United States. Congress. House. ... (Cont'd)
KF27.I549 1996l
343.73'0878--dc21
DGPO/DLC
for Library of Congress 96-217837

United States. Congress. House. Committee on International Relations.
United States. Congress. House. Committee on International Relations.
Review of the Clinton administration nonproliferation policy : hearing before the Committee on International Relations, House of Representatives, One Hundred Fourth Congress, second session, June 19, 1996. -- Washington : U.S. G.P.O. : For sale by the U.S. G.P.O., Supt. of Docs., Congressional Sales Office, 1996.
iii, 151 p. : ill. ; 24 cm.
Distributed to some depository libraries in microfiche.
Shipping list no.: 97-0090-P.
Supt. of Docs. no.: Y 4.IN 8/16:C 61/4
GPO: 1017-A-01
GPO: 1017-B-01 (MF)
ISBN 0-16-053781-9
1. Nuclear nonproliferation. 2. Nuclear arms control--Government policy--United States. 3. Arms transfers--China. 4. United States--Foreign relations--China. 5. China--Foreign relations--United States. I. Title.
KF27.I549 1996w
327.1'747'0973--dc21
DGPO/DLC
for Library of Congress 97-125419

United States. Congress. House. Committee on International Relations. Subcommittee on Asia and the Pacific.
United States. Congress. House. Committee on International Relations. Subcommittee on Asia and the Pacific.
A concurrent resolution expressing the sense of the Congress regarding a private visit by President Lee Teng-Hui of the Republic of China on Taiwan to the United States and a joint resolution relating to the United States-North Korea agreed framework and the obligations of North Korea under that and previous agreements with respect to the denuclearization of the Korean Peninsula and dialogue with the Republic of Korea : markup before the Subcommittee on Asia and the Pacific, Committee on International Relations, House of Representatives, One Hundred Fourth Congress, first session, on H.J. Res. 83 and H. Con. Res. 53, April 5, 1995.
iii, 33 p. ; 23 cm.
Distributed to some depository libraries in microfiche.

United States. Congress. House. Committee on ... (Cont'd)
United States. Congress. House. ... (Cont'd)
Shipping list no.: 97-0198-P.
Supt. of Docs. no.: Y 4.IN 8/16:R 31
GPO: 1017-A-01
GPO: 1017-B-01 (MF)
ISBN 0-16-054283-9
1. Lee, Teng-hui. 2. United States--Foreign relations--Taiwan. 3. Taiwan--Foreign relations--United States. 4. Nuclear nonproliferation. 5. United States--Foreign relations--Korea (North) 6. Korea (North)--Foreign relations--United States. I. Title.
KF27.I54915 1995m
DGPO/DLC
for Library of Congress   97-187751

United States. Congress. House. Committee on International Relations. Subcommittee on International Economic Policy and Trade.
United States. Congress. House. Committee on International Relations. Subcommittee on International Economic Policy and Trade.
  North Korean military and nuclear proliferation threat : evaluation of the U.S.-DPRK agreed framework : joint hearing before the Subcommittee on International Economic Policy and Trade and Asia and the Pacific of the Committee on International Relations, House of Representatives, One Hundred Fourth Congress, first session, February 23, 1995. -- Washington : U.S. G.P.O. : For sale by the U.S. G.P.O., Supt. of Docs., Congressional Sales Office, 1995.
    iii, 136 p. ; 24 cm.
  Distributed to some depository libraries in microfiche.
  Shipping list no.: 95-0187-P.
  Supt. of Docs. no.: Y 4.IN 8/16:K 84/5
  GPO: 1017-A-01
  GPO: 1017-B-02 (MF)
  ISBN 0-16-047140-0
  1. Nuclear nonproliferation. 2. Nuclear arms control--Korea (North) 3. United States--Military relations--Korea (North) 4. Korea (North)--Military relations--United States. I. United States. Congress. House. Committee on International Relations. Subcommittee on Asian and Pacific Affairs. II. Title.
KF27.I54924 1995h
DGPO/DLC
for Library of Congress   95-177741

United States. Congress. House. Committee on National Security.
United States. Congress. House. Committee on National Security.
Threats to U.S. national security / Committee on National Security, House of Representatives, One Hundred Fifth Congress, first session, hearing held February 13, 1997. -- Washington : U.S. G.P.O. : For sale by the U.S. G.P.O., Supt. of Docs., Congressional Sales Office, 1997.
iii, 81 p. ; 24 cm.
Distributed to some depository libraries in microfiche.
Shipping list no.: 97-0226-P.
"H.N.S.C. no. 105-11."
Supt. of Docs. no.: Y 4.SE 2/1 A:997-98/11
GPO: 1012-A-03
GPO: 1012-B-03 (MF)
ISBN 0-16-054395-9
1. National security--United States. 2. Weapons of mass destruction. 3. Nuclear nonproliferation. 4. Terrorism. 5. United States--Military policy. I. Title.
KF27.A7 1997
DGPO/DLC
for Library of Congress 97-186927

United States. Congress. House. Committee on National Security. Subcommittee on Military Procurement.
United States. Congress. House. Committee on National Security. Subcommittee on Military Procurement.
Proliferation threats and missile defense responses : hearing before the Military Procurment Subcommittee joint with Military Research and Development Subcommittee of the Committee on National Security, House of Representatives, One Hundred Fourth Congress, first and second session : hearings held April 4, 1995, February 29, March 7 and 21, June 18 and 20, September 27, 1996. -- Washington : U.S. G.P.O. : For sale by the U.S. G.P.O., Supt. of Docs., Congressional Sales Office, 1997.
v, 721 p. : ill., maps ; 24 cm.
Distributed to some depository libraries in microfiche.
Shipping list no.: 97-0259-P.
Includes bibliographical references.
"H.N.S.C. no. 104-49."
Supt. of Docs. no.: Y 4.SE 2/1 A:995-96/49
GPO: 1012-A-03
GPO: 1012-B-03 (MF)
ISBN 0-16-055007-6
1. Ballistic missile defenses--United States. 2. Nuclear nonproliferation. 3. Nuclear arms control--United States. I. United States. Congress. House. Committee on National Security. Subcommittee on

United States. Congress. House. Committee on ... (Cont'd)
United States. Congress. House. ... (Cont'd)
Military Research and Development. II. Title.
KF27.A76377 1995a
DGPO/DLC
for Library of Congress   97-186191

United States. Congress. House. Committee on Science and Technology. Subcommittee on Energy Development and Applications.
United States. Congress. House. Committee on Science and Technology. Subcommittee on Energy Development and Applications.
  Conversion of research and test reactors to low-enriched uranium (LEU) fuel : hearing before the Subcommittee on Energy Development and Applications and the Subcommittee on Energy Research and Production of the Committee on Science and Technology, U.S. House of Representatives, Ninety-eighth Congress, second session, September 25, 1984. -- Washington : U.S. G.P.O., 1985.
    xv, 1947 p. : ill. ; 24 cm.
    Distributed to some depository libraries in microfiche.
    Includes bibliographies.
    "No. 139."
    Item 1025-A-1, 1025-A-2 (microfiche)
    Supt. of Docs. no.: Y 4.Sci 2:98/139
    1. Nuclear fuels--United States. 2. Nuclear engineering--United States. 3. Uranium enrichment. 4. Nuclear nonproliferation. I. United States. Congress. House. Committee on Science and Technology. Subcommittee on Energy Research and Production. II. Title.
KF27.S3934 1984b
621.48'335--dc19
DGPO/DLC
for Library of Congress   85-601760

United States. Congress. House. Committee on Ways and Means. Subcommittee on Trade.
United States. Congress. House. Committee on Ways and Means. Subcommittee on Trade.
  United States-China trade relations : hearing before the Subcommittee on Trade of the Committee on Ways and Means, House of Representatives, One Hundred Third Congress, second session, February 24, 1994. -- Washington : U.S. G.P.O. : For sale by the U.S. G.P.O., Supt. of Docs., Congressional Sales Office, 1994.
    iv, 322 p. ; 24 cm.
    Distributed to some depository libraries in microfiche.
    Shipping list no.: 93-0468-P.
    Includes bibliographical references.
    "Serial 103-85."
    Supt. of Docs. no.: Y 4.W 36:103-85

United States. Congress. House. Committee on ... (Cont'd)
United States. Congress. House. ... (Cont'd)
  GPO: 1028-A
  GPO: 1028-B (MF)
  ISBN 0-16-046021-2
  1. Favored nation clause--United States.
  2. Reciprocity--China. 3. Human rights--China.
  4. Nuclear nonproliferation. I. Title. II. Title:
  United States China trade relations.
  KF27.W348   1994
  382'.0973051--dc20
  DGPO/DLC
  for Library of Congress   95-162305

United States. Congress. Joint Economic Committee. Subcommittee on Technology and National Security.
United States. Congress. Joint Economic Committee. Subcommittee on Technology and National Security.
    Arms trade and nonproliferation : hearings before the Subcommittee on Technology and National Security of the Joint Economic Committee, Congress of the United States, One Hundred First Congress, second session, and One Hundred Second Congress, first session. -- Washington : U.S. G.P.O. : For sale by the U.S. G.P.O., Supt. of Docs., Congressional Sales Office, 1992-
    v. <1-2   >: ill. ; 24 cm. -- (S. hrg. ; 101-1296)
    1000-B, 1000-C (MF)
    Distributed to some depository libraries in microfiche.
    Shipping list no: 92-158-P (pt. 1).
    "September 21, 1990 and April 23, 1991"--Pt. 1.
    "March 13, 1992"--Pt. 2.
    Includes bibliographical references (pt. 1, p. 288-289).
    Supt. of Docs. no.: Y 4.Ec 7:Ar 5/pt.1-
    ISBN 0-16-037370-0 (pt.1)
    1. Arms transfers--Government policy--United States.
  2. Arms control--United States. 3. Export controls--United States. 4. Nuclear nonproliferation.
  I. Title. II. Series: United States. Congress. Senate. S. hrg. ; 101-1296.
  KF25.E274   1990
  327.1'74'0973--dc20
  DGPO/DLC
  for Library of Congress   92-169300

United States. Congress. Senate. Committee on Armed Services.
United States. Congress. Senate. Committee on Armed Services.
Intelligence briefing on smuggling of nuclear material and the role of international crime organizations, and on the proliferation of cruise and ballistic missiles : hearing before the Committee on Armed Services, United States Senate, One Hundred Fourth Congress, first session, January 31, 1995. -- Washington : U.S. G.P.O. : For sale by the U.S. G.P.O., Supt. of Docs., Congressional Sales Office, 1995.
    iii, 33 p. ; 24 cm. -- (S. hrg. ; 104-35)
    Distributed to some depository libraries in microfiche.
    Shipping list no.: 95-0199-P.
    Supt. of Docs. no.: Y 4.AR 5/3:S.HRG.104-35
    GPO: 1034-A
    GPO: 1034-B (MF)
    ISBN 0-16-047202-4
    1. Illegal arms transfers. 2. Nuclear nonproliferation. 3. Arms control. I. Title. II. Series: United States. Congress. Senate. S. hrg. ; 104-35.
KF26.A7 1995d
327.1'74--dc20
DGPO/DLC
for Library of Congress   95-177771

United States. Congress. Senate. Committee on Armed Services.
United States. Congress. Senate. Committee on Armed Services.
National security implications of lowered export controls on dual-use technologies and U.S. defense capabilities : hearing before the Committee on Armed Services, United States Senate, One Hundred Fourth Congress, first session, May 11, 1995. -- Washington : U.S. G.P.O. : For sale by the U.S. G.P.O., Supt. of Docs., Congressional Sales Office, 1996.
    iii, 63 p. ; 24 cm. -- (S. hrg. ; 104-300)
    Distributed to some depository libraries in microfiche.
    Shipping list no.: 96-0130-P.
    Includes bibliographical references.
    Supt. of Docs. no.: Y 4.AR 5/3:S.HRG.104-300
    GPO: 1034-A
    GPO: 1034-B (MF)
    ISBN 0-16-052208-0
    1. Export controls--United States. 2. National security--United States. 3. Nuclear nonproliferation. 4. Arms transfers--United States. 5. Military research--United States. 6. Technology transfer--United States. I. Title. II. Series: United

United States. Congress. Senate. Committee on ... (Cont'd)
United States. Congress. Senate. ... (Cont'd)
States. Congress. Senate. S. hrg. ; 104-300.
KF26.A7 1995a
DGPO/DLC
for Library of Congress 96-120526

United States. Congress. Senate. Committee on Armed Services.
United States. Congress. Senate. Committee on Armed Services.
Security implications of the Nuclear Non-proliferation Agreement with North Korea : hearing before the Committee on Armed Services, United States Senate, One Hundred Fourth Congress, first session, January 26, 1995. -- Washington : U.S. G.P.O. : For sale by the U.S. G.P.O., Supt. of Docs., Congressional Sales Office, 1995.
iii, 95 p. : ill. ; 24 cm. -- (S. hrg. ; 104-188)
Distributed to some depository libraries in microfiche.
Shipping list no.: 96-0084-P.
Supt. of Docs. no.: Y 4.AR 5/3:S.HRG.104-188
GPO: 1034-A
GPO: 1034-B (MF)
ISBN 0-16-047732-8
1. Nuclear nonproliferation. 2. Nuclear arms control--Korea (North) 3. Nuclear weapons--Korea (North) 4. Economic assistance, American--Korea (North) 5. Korea (North)--Foreign relations--United States. 6. United States--Foreign relations--Korea (North) I. Title. II. Series: United States. Congress. Senate. S. hrg. ; 104-188.
KF26.A7 1995b
DGPO/DLC
for Library of Congress 96-128595

United States. Congress. Senate. Committee on Armed Services.
United States. Congress. Senate. Committee on Armed Services.
Worldwide threat to the United States : hearing before the Committee on Armed Services, United States Senate, One Hundred Fourth Congress, first session, January 17, 1995. -- Washington : U.S. G.P.O. : For sale by the U.S. G.P.O., Supt. of Docs., Congressional Sales Office, 1995.
iii, 93 p. ; 23 cm. -- (S. hrg. ; 104-236)
Distributed to some depository libraries in microfiche.
Shipping list no.: 96-0069-P.
Supt. of Docs. no.: Y 4.AR 5/3:S.HRG.104-236; Y 4.AR 5/3:S.HRG.104-136
GPO: 1034-A
GPO: 1034-B (MF)
ISBN 0-16-052026-6
1. Nuclear nonproliferation. 2. Nuclear terrorism.

United States. Congress. Senate. Committee on ... (Cont'd)
United States. Congress. Senate. ... (Cont'd)
3. Weapons of mass destruction. 4. Arms race.
5. Political stability. 6. National security--United States. 7. Intelligence service--United States.
8. United States--Military policy. I. Title.
II. Series: United States. Congress. Senate. S. hrg ; 104-236.
KF26.A7 1995l
DGPO/DLC
for Library of Congress   96-173836

United States. Congress. Senate. Committee on Energy and Natural Resources.
United States. Congress. Senate. Committee on Energy and Natural Resources.
 Agreement for cooperation on peaceful uses of atomic energy between the United States and the European Atomic Energy Community : hearing before the Committee on Energy and Natural Resources, United States Senate, One Hundred Third Congress, second session ... September 29, 1994. -- Washington : U.S. G.P.O. : For sale by the U.S. G.P.O., Supt. of Docs., Congressional Sales Office, 1995.
 iii, 61 p. : ill. ; 24 cm. -- (S. hrg. ; 103-944)
 Distributed to some depository libraries in microfiche.
 Shipping list no.: 95-0090-P.
 Includes bibliographical references.
 Supt. of Docs. no.: Y 4.EN 2:S.HRG.103-944
 GPO: 1040-A
 GPO: 1040-B (MF)
 ISBN 0-16-046757-8
 1. Nuclear nonproliferation. 2. Nuclear energy--Law and legislation--United States. 3. Nuclear energy--Research--International cooperation. 4. Reactor fuel reprocessing--Law and legislation--United States.
5. Euratom. I. Title. II. Series: United States. Congress. Senate. S. hrg. ; 103-944.
KF26.E551994u
DGPO/DLC
for Library of Congress   95-184809

United States. Congress. Senate. Committee on Energy and Natural Resources.
United States. Congress. Senate. Committee on Energy and Natural Resources.
 U.S.-China relations : hearing before the Committee on Energy and Natural Resources, United States Senate, One Hundred Fifth Congress, first session, on peaceful nuclear cooperation with China, October 23, 1997. -- Washington : U.S. G.P.O. : For sale by the U.S. G.P.O., Supt. of Docs., Congressional Sales Office, 1998.
 iii, 50 p. ; 24 cm. -- (S. hrg. ; 105-337)
 Distributed to some depository libraries in microfiche.

United States. Congress. Senate. Committee on ... (Cont'd)
United States. Congress. Senate. ... (Cont'd)
   Shipping list no.: 98-0161-P.
   Supt. of Docs. no.: Y 4.EN 2:S.HRG.105-337
   GPO: 1040-A
   GPO: 1040-B (MF)
   ISBN 0-16-056140-X
   1. United States--Foreign relations--China.
   2. China--Foreign relations--United States.
   3. Nuclear industry--United States. 4. Nuclear power
   plants--China. 5. Technology transfer--China.
   6. Nuclear nonproliferation. I. Title. II. Series:
   United States. Congress. Senate. S. hrg. ; 105-337.
   KF26.E55 1997m
   327.73051--dc21
   GPO/DLC
      for Library of Congress   98-143892

United States. Congress. Senate. Committee on Foreign Relations.
United States. Congress. Senate. Committee on Foreign
   Relations.
      Interpreting the Pressler amendment : commercial
   military sales to Pakistan : hearing before the
   Committee on Foreign Relations, United States Senate,
   One Hundred Second Congress, second session, July 30,
   1992. -- Washington : U.S. G.P.O. : For sale by the
   U.S. G.P.O., Supt. of Docs., Congressional Sales
   Office, 1992.
      iii, 97 p. ; 24 cm. -- (S. hrg. ; 102-859)
      Distributed to some depository libraries in
   microfiche.
      Shipping list no.: 93-0119-P.
      Includes bibliographical references.
      Supt. of Docs. no.: Y 4.F 76/2:S.HRG.102-859
   GPO: 1039-A
   GPO: 1039-B (MF)
   ISBN 0-16-039501-1
      1. Arms transfers--Pakistan. 2. Arms transfers-
   -Government policy--United States. 3. Nuclear
   nonproliferation. 4. Military assistance, American-
   -Pakistan. I. Title. II. Series: United States.
   Congress. Senate. S. hrg. ; 102-859.
   KF26.F6 1992k
   355'.0325491--dc20
   DGPO/DLC
      for Library of Congress   93-135565

United States. Congress. Senate. Committee on Foreign Relations.
United States. Congress. Senate. Committee on Foreign
   Relations.
      National security implications of missile
   proliferation : hearing before the Committee on
   Foreign Relations, United States Senate, One Hundred
   First Congress, first session, October 31, 1989. --

United States. Congress. Senate. Committee on ... (Cont'd)
United States. Congress. Senate. ... (Cont'd)
Washington : U.S. G.P.O., 1990.
iii, 88 p. : port. ; 24 cm. -- (S. hrg. ; 101-912)
1. Ballistic missiles--Government policy--United States. 2. Nuclear nonproliferation. 3. Nuclear arms control. 4. National security--United States.
I. Title. II. Series: United States. Congress. Senate. S. hrg. ; 101-912.
KF26.F6 1989j
327.1'74--dc20 90-602719
r92

United States. Congress. Senate. Committee on Foreign Relations.
United States. Congress. Senate. Committee on Foreign Relations.
Nuclear Non-Proliferation Act : joint hearing before the Committee on Foreign Relations and the Subcommittee on Energy, Nuclear Proliferation, and Government Processes of the Committee on Governmental Affairs, United States Senate, Ninety-eighth Congress, first session, September 30, 1983. -- Washington : U.S. G.P.O., 1983 [i.e. 1984]
iii, 74 p. : 1 ill. ; 24 cm. -- (S. hrg. ; 98-536)
Distributed to some depository libraries in microfiche.
Item 1039-A, 1039-B (microfiche)
Supt. of Docs. no.: Y 4.F 76/2:S.hrg.98-536
1. Nuclear nonproliferation. 2. Nuclear disarmament. I. United States. Congress. Senate. Committee on Governmental Affairs. Subcommittee on Energy, Nuclear Proliferation, and Government Processes. II. Title. III. Series: United States. Congress. Senate. S. hrg. ; 98-536.
KF26.F6 1983u
346.7304'67924--dc19
[347.306467924]
DGPO/DLC
for Library of Congress 84-601921
r902

United States. Congress. Senate. Committee on Foreign Relations.
United States. Congress. Senate. Committee on Foreign Relations.
Nuclear proliferation : learning from the Iraq experience : hearing before the Committee on Foreign Relations, United States Senate, One Hundred Second Congress, first session, October 17 and 23, 1991. -- Washington : U.S. G.P.O. : For sale by the U.S. G.P.O., Supt. of Docs., Congressional Sales Office, 1992.
iii, 58 p. : ill. ; 23 cm. -- (S. hrg. ; 102-422)
Item 1039-A, 1039-B (MF)
Distributed to some depository libraries in microfiche.

United States. Congress. Senate. Committee on ... (Cont'd)
United States. Congress. Senate. ... (Cont'd)
  Shipping list no.: 92-171-P.
  Supt. of Docs. no.: Y 4.F 76/2:S.hrg.102-422
  ISBN 0-16-037355-7
  1. Nuclear nonproliferation. 2. Nuclear arms
control--Iraq. 3. Nuclear weapons--Iraq. I. Title.
II. Series: United States. Congress. Senate. S.
hrg. ; 102-422.
  KF26.F6 1991j
  327.1'74--dc20
  DGPO/DLC
  for Library of Congress   92-172524

United States. Congress. Senate. Committee on Foreign
Relations.
United States. Congress. Senate. Committee on Foreign
Relations.
  United States-People's Republic of China nuclear
agreement : hearing before the Committee on Foreign
Relations, United States Senate, Ninety-ninth
Congress, first session, October 9, 1985. --
Washington : U.S. G.P.O., 1986.
    iii, 315 p. ; 23 cm. -- (S. hrg. ; 99-339)
    Distributed to some depository libraries in
microfiche.
    Bibliography: p. 312.
    Item 1039-A, 1039-B (microfiche)
    Supt. of Docs. no.: Y 4.F 76/2:S.hrg.99-339
    1. Nuclear nonproliferation. 2. Nuclear energy--Law
and legislation--United States. 3. Nuclear energy-
-Law and legislation--China. I. Title. II. Series:
United States. Congress. Senate. S. hrg. ; 99-339.
  KF26.F6 1985l
  341.7'34'026673051--dc19
  DGPO/DLC
  for Library of Congress   86-600941

United States. Congress. Senate. Committee on Foreign
Relations.
United States. Congress. Senate. Committee on Foreign
Relations.
  United States policy toward Iraq : human rights,
weapons proliferation, and international law : hearing
before the Committee on Foreign Relations, United
States Senate, One Hundred First Congress, second
session, June 15, 1990. -- Washington : U.S. G.P.O. :
For sale by the Supt. of Docs., Congressional Sales
Office, U.S. G.P.O., 1990.
    iii, 93 p. ; 24 cm. -- (S. hrg. ; 101-1055)
    Distributed to some depository libraries in
microfiche.
    Shipping list no.: 91-066-P.
    Item 1039-A, 1039-B (MF)
    Supt. of Docs. no.: Y 4.F 76/2:S.hrg.101-1055
    1. Human rights--Iraq. 2. Iraq--Politics and
government. 3. Nuclear nonproliferation. 4. Chemical

United States. Congress. Senate. Committee on ... (Cont'd)
United States. Congress. Senate. ... (Cont'd)
arms control. 5. Biological arms control. 6. United
States--Foreign relations--Iraq. 7. Iraq--Foreign
relations--United States. I. Title. II. Series:
United States. Congress. Senate. S. hrg. ;
101-1055.
KF26.F6 1990o
DGPO/DLC
for Library of Congress 90-603061

United States. Congress. Senate. Committee on Foreign Relations. Subcommittee on East Asian and Pacific Affairs.
United States. Congress. Senate. Committee on Foreign
Relations. Subcommittee on East Asian and Pacific
Affairs.
 Threat of North Korean nuclear proliferation :
hearings before the Subcommittee on East Asian and
Pacific Affairs of the Committee on Foreign Relations,
United States Senate, One Hundred Second Congress,
first and second sessions, November 25, 1991; January
14 and February 6, 1992. -- Washington : U.S. G.P.O. :
For sale by the U.S. G.P.O., Supt. of Docs.,
Congressional Sales Office, 1992.
 iii, 118 p. ; 24 cm. -- (S. hrg. ; 102-635)
 Item 1039-A, 1039-B (MF)
 Distributed to some depository libraries in
microfiche.
 Shipping list no.: 92-0422-P.
 Includes bibliographical references.
 Supt. of Docs. no.: Y 4.F 76/2:S.hrg.102-635
 ISBN 0-16-038687-X
 1. Nuclear weapons--Korea (North) 2. Nuclear
nonproliferation. I. Title. II. Series: United
States. Congress. Senate. S. hrg. ; 102-635.
KF26.F6354 1992
DGPO/DLC
for Library of Congress 92-209813

United States. Congress. Senate. Committee on Foreign Relations. Subcommittee on Near Eastern and South Asian Affairs.
United States. Congress. Senate. Committee on Foreign
Relations. Subcommittee on Near Eastern and South
Asian Affairs.
 Overview of U.S. policy toward South Asia : hearings
before the Subcommittee on Near Eastern and South
Asian Affairs of the Committee on Foreign Relations,
United States Senate, One Hundred Fourth Congress,
first session, March 7 and 9, 1995. -- Washington :
U.S. G.P.O. : For sale by the U.S. G.P.O., Supt. of
Docs., Congressional Sales Office, 1995.
 iii, 188 p. ; 24 cm. -- (S. hrg. ; 104-46)
 Distributed to some depository libraries in
microfiche.
 Shipping list no.: 95-0227-P.

United States. Congress. Senate. Committee on ... (Cont'd)
United States. Congress. Senate. ... (Cont'd)
Includes bibliographical references.
Supt. of Docs. no.: Y 4.F 76/2:S.HRG.104-46
GPO: 1039-A
GPO: 1039-B (MF)
ISBN 0-16-047237-7
1. United States--Foreign relations--South Asia.
2. South Asia--Foreign relations--United States.
3. United States--Foreign economic relations--South Asia.   4. South Asia--Foreign economic relations--United States.   5. Nuclear nonproliferation.
I. Title.   II. Series: United States. Congress. Senate. S. hrg. ; 104-46.
KF26.F662   1995
337.73054--dc20
DGPO/DLC
for Library of Congress   95-187780

United States. Congress. Senate. Committee on Foreign Relations. Subcommittee on Near Eastern and South Asian Affairs of the Committee on Foreign Relations, United States Senate, One Hundred Fifth Congress, first session, April 17 and May 6, 1997.
United States. Congress. Senate. Committee on Foreign Relations.   Subcommittee on Near Eastern and South Asian Affairs of the Committee on Foreign Relations, United States Senate, One Hundred Fifth Congress, first session, April 17 and May 6, 1997.
    Iran and proliferation : is the U.S. doing enough? the arming of Iran : who is responsible? : hearings before the Subcommittee on Near Eastern and South Asian Affairs of the Committee on Foreign Relations, United States Senate, One Hundred Fifth Congress, first session, April 17 and May 6, 1997. -- Washington : U.S. G.P.O. : For sale by the U.S. G.P.O., Supt. of Docs., Congressional Sales Office, 1997.
    iii, 108 p. ; 24 cm. -- (S. hrg. ; 105-289)
    "Printed for the use of the Committee on Foreign Relations."
    Includes bibliographical references.
    Supt. of Docs. no.: Y 4.F 76/2:S.HRG.105-289
GPO: 1039-A
GPO: 1039-B (MF)
ISBN 0-16-056010-1
1. Weapons of mass destruction--Iran. 2. Nuclear weapons--Iran.   3. Military assistance, Russian--Iran.
4. Military assistance, Chinese--Iran.   5. Illegal arms transfers--Iran.   6. Nuclear nonproliferation.
I. Title.   II. Series.
KF26.F662   1997
DGPO/DLC
for Library of Congress   98-143875

United States. Congress. Senate. Committee on Foreign Relations. Subcommittee on Terrorism, Narcotics, and International Operations.
United States. Congress. Senate. Committee on Foreign Relations. Subcommittee on Terrorism, Narcotics, and International Operations.
Foreign Relations authorizations for the U.S. Arms Control and Disarmament Agency : hearing before the Subcommittee on Terrorism, Narcotics, and International Operations of the Committee on Foreign Relations, United States Senate, One Hundred Second Congress, first session, March 21, 1991. -- Washington : U.S. G.P.O. : For sale by the Supt. of Docs., Congressional Sales Office, U.S. G.P.O., 1991.
iii, 62 p. : ill. ; 24 cm. -- (S. hrg. ; 102-40)
Distributed to some depository libraries in microfiche.
Shipping list no.: 91-409-P.
Item 1039-A, 1039-B (MF)
Supt. of Docs. no.: Y 4.F 76/2:S.hrg.102-40
1. United States. Arms Control and Disarmament Agency--Appropriations and expenditures. 2. Nuclear arms control. 3. Nuclear nonproliferation. I. Title. II. Series: United States. Congress. Senate. S. hrg. ; 102-40.
KF26.F685 1991
DGPO/DLC
for Library of Congress   91-601045

United States. Congress. Senate. Committee on Governmental Affairs.
United States. Congress. Senate. Committee on Governmental Affairs.
Bomb prevention vs. bomb promotion, exports in the 1990s : hearing before the Committee on Governmental Affairs, United States Senate, One Hundred Third Congress, second session, May 17, 1994. -- Washington : U.S. G.P.O. : For sale by the U.S. G.P.O., Supt. of Docs., Congressional Sales Office, 1995.
iii, 353 p. : ill. ; 24 cm. -- (S. hrg. ; 103-1028)
Distributed to some depository libraries in microfiche.
Shipping list no.: 95-0175-P.
Supt. of Docs. no.: Y 4.G 74/9:S.HRG.103-1028
GPO: 1037-B
GPO: 1037-C (MF)
ISBN 0-16-047101-X
1. Nuclear nonproliferation. 2. Nuclear arms control. 3. Atomic bomb--Materials--Marketing--Prevention. 4. Export controls--Government policy--United States. I. Title. II. Series: United States. Congress. Senate. S. hrg. ; 103-1028.

United States. Congress. Senate. Committee on ... (Cont'd)
United States. Congress. Senate. ... (Cont'd)
KF26.G67   1995c
DGPO/DLC
  for Library of Congress   95-177725

United States. Congress. Senate. Committee on Governmental
Affairs.
  United States. Congress. Senate. Committee on
Governmental Affairs.
    Disposing of plutonium in Russia : hearing before
the Committee on Governmental Affairs, United States
Senate, One Hundred Third Congress, first session,
March 9, 1993. -- Washington : U.S. G.P.O. : For sale
by the U.S. G.P.O., Supt. of Docs., Congressional
Sales Office, 1993.
    iii, 194 p. : ill. ; 24 cm. -- (S. hrg. ; 103-135)
    Distributed to some depository libraries in
microfiche.
    Shipping list no.: 93-0527-P.
    Includes bibliographical references.
    Supt. of Docs. no.: Y 4.G 74/9:S.HRG.103-135
    GPO: 1037-B
    GPO: 1037-C (MF)
    ISBN 0-16-041263-3
    1. Plutonium. 2. Nuclear nonproliferation.
3. Explosive ordnance disposal. 4. Arms control-
-Russia (Federation) I. Title. II. Series: United
States. Congress. Senate. S. hrg. ; 103-135.
KF26.G67   1993a
DGPO/DLC
  for Library of Congress   93-245890

United States. Congress. Senate. Committee on Governmental
Affairs.
  United States. Congress. Senate. Committee on
Governmental Affairs.
    Nuclear and missile proliferation : hearing before
the Committee on Governmental Affairs, United States
Senate, One Hundred First Congress, first session, May
18, 1989. -- Washington : U.S. G.P.O. : For sale by
the Supt. of Docs., Congressional Sales Office, U.S.
G.P.O., 1990.
    iii, 90 p. : ill., maps ; 24 cm. -- (S. hrg. ;
101-562)
    Distributed to some depository libraries in
microfiche.
    Shipping list no.: 90-263-P.
    Item 1037-B, 1037-C (MF)
    Supt. of Docs. no.: Y 4.G 74/9:S.hrg.101-562
    1. Nuclear nonproliferation. 2. Nuclear arms
control. 3. Ballistic missiles. 4. United States.-
-Defenses. I. Title. II. Series: United States.
Congress. Senate. S. hrg. ; 101-562.

United States. Congress. Senate. Committee on ... (Cont'd)
United States. Congress. Senate. ... (Cont'd)
KF26.G67 1989y
327.1'74--dc20
DGPO/DLC
for Library of Congress   90-600994

United States. Congress. Senate. Committee on Governmental Affairs.
United States. Congress. Senate. Committee on Governmental Affairs.
   Nuclear non-proliferation : hearing before the Committee on Governmental Affairs, United States Senate, One Hundred Fourth Congress, first session, March 14, 1995. -- Washington : U.S. G.P.O. : For sale by the U.S. G.P.O., Supt. of Docs., Congressional Sales Office, 1996.
   iii, 40 p. ; 24 cm. -- (S. hrg. ; 104-519)
   Distributed to some depository libraries in microfiche.
   Shipping list no.: 96-0338-P.
   Supt. of Docs. no.: Y 4.G 74/9:S.HRG.104-519
   GPO: 1037-B
   GPO: 1037-C (MF)
   ISBN 0-16-052948-4
   1. Nuclear nonproliferation. 2. Treaty on the Non-proliferation of Nuclear Weapons (1968)   I. Title. II. Series: United States. Congress. Senate. S. hrg. ; 104-519.
KF26.G67 1995u
DGPO/DLC
for Library of Congress   97-118007

United States. Congress. Senate. Committee on Governmental Affairs.
United States. Congress. Senate. Committee on Governmental Affairs.
   Nuclear non-proliferation and U.S. national security : hearings before the Committee on Governmental Affairs, United States Senate, One hundredth Congress, first session, February 24, 25, and March 5, 1987. -- Washington : U.S. G.P.O. : For sale by the Supt. of Docs., Congressional Sales Office, U.S. G.P.O., 1987.
   iv, 225 p. : ill., 2 maps ; 24 cm. -- (S. hrg. ; 100-88)
   Distributed to some depository libraries in microfiche.
   Shipping list no.: 87-457-P.
   Bibliography: p. 111.
   Item 1037-B, 1037-C (microfiche)
   Supt. of Docs. no.: Y 4.G 74/9:S.hrg.100-88
   1. Nuclear nonproliferation. 2. National security--United States.   I. Title.   II. Title: Nuclear non-proliferation and US national security.   III. Series: United States. Congress. Senate. S. hrg. ; 100-88.

United States. Congress. Senate. Committee on ... (Cont'd)
United States. Congress. Senate. ... (Cont'd)
KF26.G67   1987b
327.1'74--dc19
DGPO/DLC
for Library of Congress   87-602062
                                                r92

United States. Congress. Senate. Committee on Governmental
Affairs.
  United States. Congress. Senate. Committee on
Governmental Affairs.
    Proliferation and regional security in the 1990's :
hearing before the Committee on Governmental Affairs,
United States Senate, One Hundred First Congress,
second session, October 9, 1990. -- Washington : U.S.
G.P.O. : For sale by the Supt. of Docs., Congressional
Sales Office, U.S. G.P.O., 1991.
    iii, 88 p. ; 24 cm. -- (S. hrg. ; 101-1208)
    1. Nuclear nonproliferation. 2. National security--Middle East. 3. Nuclear weapons--Government policy--United States. I. Title. II. Series: United
States. Congress. Senate. S. hrg. ; 101-1208.
KF26.G67   1990ag   91-600682
                                                r92

United States. Congress. Senate. Committee on Governmental
Affairs.
  United States. Congress. Senate. Committee on
Governmental Affairs.
    Proliferation threats of the 1990's : hearing before
the Committee on Governmental Affairs, United States
Senate, One Hundred Third Congress, first session,
February 24, 1993. -- Washington : U.S. G.P.O. : For
sale by the U.S. G.P.O., Supt. of Docs., Congressional
Sales Office, 1993.
    iii, 192 p. : ill. ; 24 cm. -- (S. hrg. ; 103-208)
    Distributed to some depository libraries in
microfiche.
    Shipping list no.: 93-0578-P.
    Includes bibliographical references.
    Supt. of Docs. no.: Y 4.G 74/9:S.HRG.103-208
    GPO: 1037-B
    GPO: 1037-C (MF)
    ISBN 0-16-041553-5
    1. Nuclear nonproliferation. 2. Nuclear arms
control. 3. Biological arms control. 4. Chemical
arms control.   I. Title. II. Series: United States.
Congress. Senate. S. hrg. ; 103-208.
KF26.G67   1993
DGPO/DLC
for Library of Congress   93-230665

United States. Congress. Senate. Committee on Governmental Affairs.
United States. Congress. Senate. Committee on Governmental Affairs.
U.S.-EURATOM agreement for peaceful nuclear cooperation : hearing before the Committee on Governmental Affairs, United States Senate, One Hundred Fourth Congress, second session, February 28, 1996. -- Washington : U.S. G.P.O. : For sale by the U.S. G.P.O., Supt. of Docs., Congressional Sales Office, 1996.
iv, 443 p. : ill., map ; 24 cm. -- (S. hrg. ; 104-481)
Distributed to some depository libraries in microfiche.
Shipping list no.: 96-0303-P.
Includes bibliographical references.
Supt. of Docs. no.: Y 4.G 74/9:S.HRG.104-481
GPO: 1037-B
GPO: 1037-C (MF)
ISBN 0-16-052797-X
1. Euratom--United States. 2. Nuclear nonproliferation--International cooperation. 3. Nuclear industry--Government policy--United States. 4. Nuclear industry--Government policy--Europe. 5. Export controls--United States. 6. Export controls--Europe. I. Title. II. Series: United States. Congress. Senate. S. hrg. ; 104-481.
KF26.G67 1996a
327.1'747'091821--dc21
DGPO/DLC
for Library of Congress   96-202517

United States. Congress. Senate. Committee on Governmental Affairs.
United States. Congress. Senate. Committee on Governmental Affairs.
Weapons proliferation in the new world order : hearing before the Committee on Governmental Affairs, United States Senate, One Hundred Second Congress, second session, January 15, 1992. -- Washington : U.S. G.P.O. : For sale by the U.S. G.P.O., Supt. of Docs., Congressional Sales Office, 1992.
iii, 43 p. ; 23 cm. -- (S. hrg. ; 102-720)
Item 1037-B, 1037-C (MF)
Distributed to some depository libraries in microfiche.
Shipping list no.: 92-0593-P.
Supt. of Docs. no.: Y 4.G 74/9:S.hrg.102-720
ISBN 0-16-039043-5
1. Arms race. 2. Arms control. 3. Arms control--Former Soviet republics. 4. Nuclear arms control. 5. Nuclear arms control--Former Soviet republics. 6. Nuclear nonproliferation. I. Title. II. Series: United States. Congress. Senate. S. hrg. ; 102-720.

United States. Congress. Senate. Committee on ... (Cont'd)
United States. Congress. Senate. ... (Cont'd)
KF26.G67   1992
327.1'74--dc20
DGPO/DLC
for Library of Congress   92-240385

United States. Congress. Senate. Committee on Governmental
Affairs. Subcommittee on Energy, Nuclear Proliferation,
and Government Processes.
   United States. Congress. Senate. Committee on
Governmental Affairs. Subcommittee on Energy, Nuclear
Proliferation, and Government Processes.
   Review of 1985 U.S. government nonproliferation
activities : hearing before the Subcommittee on
Energy, Nuclear Proliferation, and Government
Processes of the Committee on Governmental Affairs,
United States Senate, Ninety-ninth Congress, second
session, April 10, 1986. -- Washington : U.S. G.P.O. :
For sale by the Supt. of Docs., Congressional Sales
Office, U.S. G.P.O., 1986.
   iii, 62 p. ; 24 cm. -- (S. hrg. ; 99-668)
   Distributed to some depository libraries in
microfiche.
   Shipping list no.: 86-644-P.
   Item 1037-B, 1037-C (microfiche)
   Supt. of Docs. no.: Y 4.G 74/9:S.hrg.99-668
   1. Nuclear nonproliferation. I. Title. II. Title:
Review of 1985 US government nonproliferation
activities. III. Series: United States. Congress.
Senate. S. hrg. ; 99-668.
KF26.G6729   1986a
327.1'74--dc19
DGPO/DLC
for Library of Congress   86-602608

United States. Congress. Senate. Committee on Governmental
Affairs. Subcommittee on Energy, Nuclear Proliferation,
and Government Processes.
   United States. Congress. Senate. Committee on
Governmental Affairs. Subcommittee on Energy, Nuclear
Proliferation, and Government Processes.
   Third Non-Proliferation Treaty Review Conference and
29th regular session of the General Conference of the
International Atomic Energy Agency : hearing before
the Subcommittee on Energy, Nuclear Proliferation, and
Government Processes of the Committee on Governmental
Affairs, United States Senate, Ninety-ninth Congress,
first session, November 20, 1985. -- Washington : U.S.
G.P.O., 1986.
   iii, 99 p. ; 24 cm. -- (S. hrg. ; 99-488)
   Distributed to some depository libraries in
microfiche.
   Shipping list no.: 86-315-P.
   Item 1037-B, 1037-C (microfiche)
   Supt. of Docs. no.: Y 4.G 74/9:S.hrg.99-488
   1. Nuclear nonproliferation--Congresses. 2. Non

United States. Congress. Senate. Committee on ... (Cont'd)
United States. Congress. Senate. ... (Cont'd)
-Proliferation Treaty Review Conference (3rd : 1985 : Geneva, Switzerland) 3. International Atomic Energy Agency. General Conference (29th : 1985 : Vienna, Austria) I. Title. II. Series: United States. Congress. Senate. S. hrg. ; 99-488.
KF26.G6729 1985d
327.1'74--dc19
DGPO/DLC
for Library of Congress   86-601496

United States. Congress. Senate. Committee on Governmental Affairs. Subcommittee on Energy, Nuclear Proliferation, and Government Processes.
United States. Congress. Senate. Committee on Governmental Affairs. Subcommittee on Energy, Nuclear Proliferation, and Government Processes.
    United Nations Association report on nuclear nonproliferation : hearing before the Subcommittee on Energy, Nuclear Proliferation, and Government Processes of the Committee on Governmental Affairs, United States Senate, Ninety-eighth Congress, second session, June 27, 1984. -- Washington : U.S. G.P.O., 1984.
    iii, 87 p. : ill. ; 24 cm. -- (S. hrg. ; 98-909)
    Distributed to some depository libraries in microfiche.
    Bibliography: p. 87.
    Item 1037-B, 1037-C (microfiche)
    Supt. of Docs. no.: Y 4.G 74/9:S.hrg.98-909
    1. Nuclear nonproliferation. 2. United Nations Association of the United States of America.
I. Title. II. Series: United States. Congress. Senate. S. hrg. ; 98-909.
KF26.G6729 1984a
327.1'74--dc19
DGPO/DLC
for Library of Congress   84-603519
                                                                 r89

United States. Congress. Senate. Committee on Governmental Affairs. Subcommittee on Federal Services, Post Office, and Civil Service.
United States. Congress. Senate. Committee on Governmental Affairs. Subcommittee on Federal Services, Post Office, and Civil Service.
    A review of arms export licensing : hearing before the Subcommittee on Federal Services, Post Office, and Civil Service of the Committee on Governmental Affairs, United States Senate, One Hundred Third Congress, second session, June 15, 1994. -- Washington : U.S. G.P.O. : For sale by the U.S. G.P.O., Supt. of Docs., Congressional Sales Office, 1994.
    iii, 163 p. : ill. ; 23 cm. -- (S. hrg. ; 103-670)
    Distributed to some depository libraries in microfiche.

United States. Congress. Senate. Committee on ... (Cont'd)
United States. Congress. Senate. ... (Cont'd)
Shipping list no.: 94-0281-P.
Includes bibliographical references.
Supt. of Docs. no.: Y 4.G 74/9:S.HRG.103-670
GPO: 1037-B
GPO: 1037-C (MF)
ISBN 0-16-044780-1
1. Export controls--United States. 2. Arms control--United States. 3. Nuclear nonproliferation.
I. Title. II. Series: United States. Congress. Senate. S. hrg. ; 103-670.
KF26.G67313   1994a
DGPO/DLC
for Library of Congress   94-232660

United States. Congress. Senate. Committee on Governmental Affairs. Subcommittee on International Security, Proliferation, and Federal Services.
United States. Congress. Senate. Committee on Governmental Affairs. Subcommittee on International Security, Proliferation, and Federal Services.
  Proliferation and U.S. export controls : hearing before the Subcommittee on International Security, Proliferation, and Federal Services of the Committee on Governmental Affairs, United States Senate, One Hundred Fifth Congress, first session, June 11, 1997. -- Washington : U.S. G.P.O. : For sale by the U.S. G.P.O., Supt. of Docs., Congressional Sales Office, 1997.
  iii, 53 p. ; 24 cm. -- (S. hrg. ; 105-238)
  Distributed to some depository libraries in microfiche.
Shipping list no.: 98-0096-P.
Supt. of Docs. no.: Y 4.G 74/9:S.HRG.105-238
GPO: 1037-B
GPO: 1037-C (MF)
ISBN 0-16-055864-6
1. Export controls--United States.
2. Supercomputers--United States. 3. Technology transfer--China. 4. Nuclear nonproliferation.
I. Title. II. Series: United States. Congress. Senate. S. hrg. ; 105-238.
KF26.G6739   1997a
GPO/DLC
for Library of Congress   98-139342

United States. Congress. Senate. Committee on Governmental Affairs. Subcommittee on International Security, Proliferation, and Federal Services.
United States. Congress. Senate. Committee on Governmental Affairs. Subcommittee on International Security, Proliferation, and Federal Services.
  Proliferation--Russian case studies : hearing before the Subcommittee on International Security, Proliferation, and Federal Services of the Committee on Governmental Affairs, United States Senate, One Hundred Fifth Congress, first session, June 5, 1997. -- Washington : U.S. G.P.O. : For sale by the U.S. G.P.O., Supt. of Docs., Congressional Sales Office, 1997.
    iii, 47 p. ; 23 cm. -- (S. hrg. ; 105-237)
  Distributed to some depository libraries in microfiche.
  Shipping list no.: 98-0096-P.
  Includes bibliographical references.
  Supt. of Docs. no.: Y 4.G 74/9:S.HRG.105-237
  GPO: 1037-B
  GPO: 1037-C (MF)
  ISBN 0-16-055868-9
  1. Nuclear weapons industry--Russia (Federation)
  2. Weapons of mass destruction--Russia (Federation)
  3. Illegal arms transfers--Russia (Federation)
  4. Military assistance, Russian--Iran. 5. Nuclear nonproliferation. I. Title. II. Series: United States. Congress. Senate. S. hrg. ; 105-237.
  KF26.G6739 1997e
  GPO/DLC
  for Library of Congress   98-139070

United States. General Accounting Office.
United States. General Accounting Office.
  Hong Kong's reversion to China : effective monitoring critical to assess U.S. nonproliferation risks : report to congressional requesters / United States General Accounting Office. -- Washington, D.C. (Room 1100, 700 4th St. NW, Washington 20548-0001) : GAO, [1997]
    49 p. : ill. ; 28 cm.
  Cover title.
  "May 1997."
  "GAO/NSIAD-97-149."
  "B-275463"--P. 1.
  1. Nuclear nonproliferation. 2. Export controls--United States. 3. Hong Kong (China)--History--Transfer of Sovereignty from Great Britain, 1997.
  I. Title.
  JZ5675.U557 1997
  327.1'747--dc21   97-187715

United States. General Accounting Office.
United States. General Accounting Office.
Nuclear nonproliferation : export licensing procedures for dual-use items need to be strengthened : report to the chairman, Committee on Governmental Affairs, U.S. Senate / United States General Accounting Office. -- Washington, D.C. : The Office, [1994]
   69 p. : ill. ; 28 cm.
   Cover title.
   "April 1994."
   "GAO/NSIAD-94-119."
   "B-256585"--P. [1].
   Supt. of Docs. no.: GA 1.13:NSIAD-94-119
   1. Export controls--United States--Evaluation. 2. Nuclear nonproliferation. I. Title.
   JX1974.73.U54   1994
   382'.4562345119'0973--dc20   94-190829

United States. General Accounting Office.
United States. General Accounting Office.
Nuclear nonproliferation : U.S. international nuclear materials tracking capabilities are limited : report to Congressional requesters / United States General Accounting Office. -- Washington, D.C. : The Office, [1994]
   27 p. : ill. ; 28 cm.
   Cover title.
   "December 1994."
   "GAO/RCED/AIMD-95-5."
   "B-259533"--P. 1.
   Supt. of Docs. no.: GA 1.13:RCED/AIMD-95-5
   1. Nuclear nonproliferation. 2. Nuclear arms control--United States. 3. United States. Dept. of Energy--Rules and practice. I. Title.
   JX1974.73.U54   1994a   95-143444

United States. General Accounting Office.
United States. General Accounting Office.
   Nuclear nonproliferation : information on nuclear exports controlled by U.S.-EURATOM agreement : report to the Committee on Governmental Affairs, U.S. Senate / United States General Accounting Office. -- Washington, D.C. : The Office, [1995]
   45 p. : ill., map ; 28 cm.
   Cover title.
   "June 1995."
   "GAO/RCED-95-168."
   "B-261275"--P. 1.
   Supt. of Docs. no.: GA 1.13:RCED-95-168
   1. Nuclear nonproliferation. 2. Export controls--United States. I. United States. Congress. Senate. Committee on Governmental Affairs. II. Title.

United States. General Accounting Office. (Cont'd)
United States. General Accounting Office. (Cont'd)
JX1974.73.U54   1995   95-188975
United States. General Accounting Office.
United States. General Accounting Office.
   Nuclear nonproliferation : implications of the
U.S./North Korean agreement on nuclear issues : report
to the Chairman, Committee on Energy and Natural
Resources, U.S. Senate / United States General
Accounting Office. -- Washington, D.C. : The Office ;
Gaithersburg, MD (P.O. Box 6015, Gaithersburg
20884-6015) : The Office [distributor, 1996]
   63 p. : map ; 28 cm.
   Cover title.
   "October 1996."
   Includes bibliographical references.
   "GAO/RCED/NSIAD-97-8."
   "B-272530"--P. 1.
   Supt. of Docs. no.: GA 1.13:RCED/NSIAD-97-8
   1. Nuclear nonproliferation. 2. Nuclear arms
control--Korea (North) 3. Nuclear reactors--Korea
(North) 4. Economic assistance, American--Korea
(North) 5. United States--Foreign relations--Korea
(North) 6. Korea (North)--Foreign relations--United
States. I. United States. Congress. Senate.
Committee on Energy and Natural Resources. II. Title.
JZ5675.U56   1996
341.7'34--dc21   96-223634

United States. General Accounting Office.
United States. General Accounting Office.
   Nuclear nonproliferation : implementation of the
U.S./North Korean agreed framework on nuclear issues :
report to the Chairman, Committee on Energy and
Natural Resources, U.S. Senate / United States General
Accounting Office. -- Washington, D.C. (Room 1100, 700
4th St. NW, Washington 20548-0001) : GAO, [1997]
   66 p. : ill. ; 28 cm.
   Cover title.
   "June 1997."
   Includes bibliographical references.
   "GAO/RCED/NSIAD-97-165."
   "B-276968"--P. 1.
   1. Nuclear nonproliferation. 2. Nuclear arms
control--Korea (North) 3. United States--Foreign
relations--Korea (North) 4. Korea (North)--Foreign
relations--United States. I. United States.
Congress. Senate. Committee on Energy and Natural
Resources. II. Title.
JZ5675.U559   1997
327.1'743--dc21   97-188868

United States. President (1989-1993 : Bush)
United States. President (1989-1993 : Bush)
Activities to prevent nuclear proliferation : communication from the President of the United States transmitting his annual report, reviewing all activities of U.S. Government Departments and Agencies during calendar 1990, relating to the prevention of nuclear proliferation, pursuant to 22 U.S.C. 3281. -- Washington : U.S. G.P.O., 1991.
80 p. ; 24 cm. -- (House document ; 102-135)
1. Nuclear nonproliferation. 2. Nuclear arms control--United States. I. Series: House document (United States. Congress. House) ; no. 102-135.
JX1974.73.A4   1991
327.1'74'0973--dc20
DGPO/DLC
for Library of Congress   93-135239

United States. General Accounting Office.
United States. General Accounting Office.
Nuclear nonproliferation : status of U.S. efforts to improve nuclear material controls in newly independent states / United States General Accounting Office. -- Washington, D.C. : The Office, [1996]
46 p. : ill. ; 28 cm.
Cover title.
"March 1996."
"GAO/NSIAD/RCED-96-89."
Includes bibliographical references (p.46).
"B-270052"--P. [1].
1. Nuclear nonproliferation. 2. Technical assistance, American--Former Soviet republics. 3. Weapons of mass destruction. I. Title.
JX1974.73.U54   1996
327.1'747'0947--dc21   96-140247

Van Creveld, Martin L.
Van Creveld, Martin L.
Nuclear proliferation and the future of conflict / Martin van Creveld. -- New York : Free Press ; Toronto : Maxwell Macmillan Canada ; New York : Maxwell Macmillan International, c1993.
viii, 180 p. ; 25 cm.
Includes bibliographical references (p. 161-173) and index.
ISBN 0-02-933156-0 : $22.95
1. Nuclear nonproliferation. 2. War (International law) 3. Military policy. I. Title.
JX1974.73.V36   1993
355.02'17--dc20   93-212

Vance, Mary A.
Vance, Mary A.
   Nuclear nonproliferation : a bibliography / Mary Vance. -- Monticello, Ill., USA : Vance Bibliographies, [1989]
   20 p. ; 28 cm. -- (Public administration series--bibliography, ISSN 0193-970X ; P 2763)
   Cover title.
   "November 1989."
   ISBN 0-7920-0353-5 : $5.00
   1. Nuclear nonproliferation--Bibliography.
   I. Title. II. Series.
   Z6464.D6V35   1989
   [JX1974.73]
   016.3271'74--dc20   90-113607

Wendt, James C., 1944-
Wendt, James C., 1944-
   The North Korean nuclear program : what is to be done? / James C. Wendt. -- Santa Monica : Rand, c1994.
   xii, 27 p. ; 28 cm.
   "Prepared for the U.S. Army."
   "Arroyo Center."
   Includes bibliographical references.
   ISBN 0-8330-1534-6
   1. United States--Military policy. 2. Nuclear weapons--Korea (North) 3. Nuclear nonproliferation. I. United States. Army. II. Arroyo Center. III. Title.
   UA23.W374   1994
   327.1'74'095193--dc20   94-10797
                                           r94

Whitmore, D. C.
Whitmore, D. C.
   Characterization of the nuclear proliferation threat / D.C. Whitmore. -- Auburn, WA (P.O. Box 1105, Auburn 98071-1105) : D.C. Whitmore, c1993.
   100 p. : ill. ; 29 cm. -- (Monograph series on security and arms control)
   "April 1993."
   Includes bibliographical references (p. 94-100).
   1. Nuclear nonproliferation. 2. Security, International. 3. United States--Military policy.
   I. Title. II. Series.
   JX1974.73.W48   1993
   327.1'74--dc20   94-221310

Wilkening, Dean, 1950-
Wilkening, Dean, 1950-
   Nuclear deterrence in a regional context / Dean Wilkening, Kenneth Watman. -- Santa Monica, CA : Rand, 1995.
   xv, 75 p. : ill. ; 23 cm.
   "Prepared for the United States Army and United

Wilkening, Dean, 1950- (Cont'd)
Wilkening, Dean, 1950- (Cont'd)
States Air Force.
"Arroyo Center."
"Project Air Force."
"Rand."
Includes bibliographical references (p. 71-75).
ISBN 0-8330-1596-6
1. United States--Military policy. 2. Deterrence (Strategy) 3. Nuclear nonproliferation. I. Watman, Kenneth, 1948- . II. United States. Army. III. United States. Air Force. IV. Arroyo Center. V. Project Air Force (U.S.) VI. Rand Corporation. VII. Title.
UA23.W4582 1995
355.02'17--dc20   94-36395
                                                    r982

Wilson, Michael.
Wilson, Michael.
The nuclear future : Asia and Australia and the 1995 Conference on Non-Proliferation / Michael Wilson. -- Queensland : Centre for the Study of Australia-Asia Relations, 1995.
55 p. ; 25 cm. -- (Australia-Asia paper ; no. 74)
Includes bibliographical references.
ISBN (invalid) 086857595x
1. Nuclear nonproliferation. 2. Australia--Foreign relations--1945- I. Griffith University. Centre for the Study of Australian-Asian Relations. II. Title. III. Series.
JX1974.73.W55 1995   96-139051

Workshop on the Technology for Arms Control Verification in the 1990s (1991 : Toronto, Ont.)
Workshop on the Technology for Arms Control Verification in the 1990s (1991 : Toronto, Ont.)
Controlling the global arms threat : proceedings of a Workshop on the Technology for Arms Control Verification in the 1990s, IEEE Canada/Science for Peace, Ryerson Polytechnical Institute, Toronto, 20 June 1991 / edited by Peter Brogden and Walter Dorn. -- Ottawa : Canadian Centre for Arms Control and Disarmament = Centre canadien pour le contrôle des armements et le désarmement, 1992.
viii, 102 p. : ill. ; 28 cm. -- (Aurora papers, ISSN 0825-1916 ; 12)
C92-90424-6
Includes bibliographical references.
ISBN 0-920357-25-3 : $12.95
1. Arms control--Verification--Congresses. 2. Disarmament--Inspection--Congresses. 3. Military surveillance--Congresses. 4. Nuclear nonproliferation--Congresses. I. Brogden, Peter, 1934- . II. Dorn, Walter H., 1961- .
III. Canadian Centre for Arms Control and Disarmament. IV. Institute of Electrical and Electronics Engineers.

Workshop on the Technology for Arms Control ... (Cont'd)
Workshop on the Technology for Arms Control ... (Cont'd)
V. Science for Peace (Association) VI. Ryerson
Polytechnical Institute. VII. Series.
UA12.5.W67 1992
327.1'74--dc20					93-134516

Zinberg, Dorothy S.
Zinberg, Dorothy S.
    The missing link? : nuclear proliferation and the
international mobility of Russian nuclear experts /
Dorothy S. Zinberg. -- New York : United Nations,
1995.
    vi, 46 p. ; 21 cm. -- (Research papers / United
Nations Institute for Disarmament Research, ISSN
1014-4013 ; no. 35)
    "UNIDIR/95/27."
    "United Nations publication sales no.
GV.E.95.0.18."--T.p. verso.
    Includes bibliographical references.
    UNIDIR/95/27; undocs
    ISBN 9290451041
    1. Nuclear nonproliferation. 2. Scientists in
government--Russia (Federation). 3. Brain drain-
-Russia (Federation) I. United Nations Institute for
Disarmament Research. II. Title. III. Series:
Research paper (United Nations Institute for
Disarmament Research) ; no. 35.
KZ5675.Z56 1995					96-216399

Part 2
Journals
1994-Present (1999)

```
*************************************************************************
FILE:           CRS PUBLIC POLICY LITERATURE (PPLT)     July 31, 1998

                Citations are arranged by year with the most current
                appearing first.
*************************************************************************

TITLE:          Rethinking the nuclear equation: the United States and the
                new nuclear powers.
SOURCE:         Washington quarterly, v. 17, winter 1994: 5-25.
AUTHOR:         Dunn, Lewis A.
LOCATION:       LRS94-2                                         UC 650 A

NOTES:          "In the proliferation arena, preventing the further spread
                of nuclear weapons should remain the top U.S. priority.
                For too long, however, U.S. policymakers and defense
                planners have neglected the need to focus seriously on how
                to deal with the consequences of proliferation for American
                security. Faced with the emergence of several undeclared
                nuclear powers, with others on the horizon, and with the
                loss of cold war missions, counterproliferation is now the
                vogue."

SUBJECT(S):     Nuclear nonproliferation / Collective security / Nuclear
                weapons
```

```
TITLE:          North Korea.
SOURCE:         Far Eastern economic review, v. 157, Feb. 10, 1994: 16-20,
                22-23.
LOCATION:       LRS94-1949                                      DS 930 E

NOTES:          Contents.--Bomb and bombast, by Nayan Chanda.--Hide and
                seek, by Nayan Chanda.--Political radiation, by Shim Jae
                Hoon.--Prepared for the worst, by Ed Paisley.

SUBJECT(S):     Nuclear weapons--[North Korea] / Arms control verification-
                -[North Korea] / Arms control / Nuclear nonproliferation--
                [North Korea]
```

```
TITLE:          Prospects for Ukrainian denuclearization after the Moscow
                trilateral statement.
SOURCE:         Arms control today, v. 24, Mar. 1994: 21-26.
LOCATION:       LRS94-2524                                      DK 508

NOTES:          "Arms Control Association (ACA) held a news conference on
                January 28 to provide background and context on the many
                issues that affect prospects for Ukrainian
                denuclearization. Panelists included Spurgeon M. Keeny,
                Jr., Jack Mendelsohn, F. Stephen Larrabee, and Raymond
                Garthoff."
```

SUBJECT(S): Nuclear weapons--[Ukraine]--Addresses, statements, etc. / Nuclear nonproliferation--[Ukraine]--Addresses, statements, etc. / Arms control

TITLE: A nuclear North Korea: the choices are narrowing.
SOURCE: World policy journal, v. 11, summer 1994: 27-35.
AUTHOR: Mack, Andrew.
LOCATION: LRS94-5426
DS 930 E

NOTES: "Pyongyang may be playing a delicate and dangerous game of brinkmanship and still probing to see if it can extract any further concessions from Washington."

SUBJECT(S): Nuclear weapons--[North Korea] / Arms control verification--[North Korea] / Nuclear facilities--[North Korea] / Nuclear nonproliferation--[North Korea]

TITLE: Nuclear non-proliferation and the international system.
SOURCE: Contemporary review, v. 268, Mar. 1996: 134-140.
AUTHOR: Tripodi, Paolo.
LOCATION: LRS96-3580
UC 650 A

NOTES: "The dissolution of the Warsaw Pact and the Soviet Union collapse have deeply changed the geo-strategical outlook. The Cold War was unexpectedly over and the role of NATO itself was being seriously put into question. Within this new and more complex situation, a trend to non-proliferation emerged, that is elimination of the nuclear threat by scaling down and eventually eliminating all nuclear weapons."

SUBJECT(S): Nuclear nonproliferation / Nuclear weapons / Arms control / International relations / Treaty on the Nonproliferation of Nuclear Weapons (NPT)

TITLE: Capping Israel's nuclear volcano.
SOURCE: Israel affairs, v. 2, autumn 1995: 93-111.
AUTHOR: Barnaby, Frank.
LOCATION: LRS95-13672
Optical Disk

NOTES: "Brief survey of the nuclear capabilities of Israel, Iran and Pakistan .... Nuclear weapons are essential to guarantee Israel's existence. This makes it politically impossible for the Israeli leaderhip to agree to give up nuclear weapons."

SUBJECT(S): Nuclear weapons--[Israel] / Nuclear research--[Israel] / Nuclear nonproliferation--[Middle East] / Arms control--[Middle East] / National security--[Israel] / Nuclear weapons--[Middle East] / Iran / Israel / Pakistan

TITLE:	Middle East nuclear isues in global perspective.
SOURCE:	Middle East policy, v. 4, Sept. 1995: 188-209.
AUTHOR:	Power, Paul F.
LOCATION:	LRS95-13675 Optical Disk
NOTES:	"There are two immediate questions on the region's nuclear agenda. The first concerns the prospects for additional bombspread by a 'rogue' . . . . The second question focuses on the odds for rolling back or at least capping existing regional proliferation . . . . This paper offers an analysis of selected topics which bear on the two questions."
SUBJECT(S):	Nuclear nonproliferation--[Middle East] / Nuclear weapons--[Middle East] / Arms control--[Middle East] / Libya / Iraq / Iran

TITLE:	Department of Energy; poor management of nuclear materials tracking system makes success unlikely; report to the Ranking Minority Member, Committee on Government Affairs, U.S. Senate. Aug. 3, 1995.
SOURCE:	Washington, G.A.O., 1995. 15 p.
AUTHOR:	U.S. General Accounting Office.
LOCATION:	LRS95-13871 AVAILABLE FROM ISSUING AGENCY Optical Disk
NOTES:	"GAO/AIMD-95-165, B-260569" The Nuclear Materials Management and Safeguards Systems (NMMSS) "is the United States' official nuclear materials tracking and accounting system. NMMSS provides information on nuclear materials to support both domestic programs and international nuclear policies . . . . Tracking and accounting for the hundreds of tons of plutonium, highly enriched uranium, and other nuclear materials that have accumulated are important to help (1) ensure that nuclear materials are used only for peaceful purposes, (2) protect nuclear materials from loss, theft, or other diversion, (3) comply with international treaty obligations, and (4) provide data to policymakers and other government officials on the amount and location of nuclear materials."
SUBJECT(S):	Nuclear security measures--[U.S.] / Nuclear nonproliferation--[U.S.] / Nuclear exports / Plutonium / Uranium / U.S. Dept. of Energy

TITLE:	Challenges & opportunities: U.S. nonproliferation and counterproliferation programs in 1996.
SOURCE:	Washington, Center for Strategic and Budgetary Assessments, 1996. 38 p.
AUTHOR:	Kosiak, Steven M.
LOCATION:	LRS96-4391 UA 17

NOTES: "The potential dangers posed by the proliferation of 'weapons of mass destruction' (WMD) and the means to deliver them." Contents.--International atomic energy agency (IAEA) safeguards program.--Nunn-Lugar cooperative threat reduction (CTR) program.--Chemical and biologial warfare (CBW) defense programs.--Ballistic missile defense (BMD) programs.--Precision-strike systems.

SUBJECT(S): Nuclear nonproliferation / Arms control--[U.S.] / Weapons of mass destruction
ADDED ENTRY: Center for Strategic and Budgetary Assessments.

TITLE: U.S.-Euratom agreement for peaceful nuclear cooperation. Feb. 28, 1996.
SOURCE: Washington, G.P.O., 1996. 443 p. (Hearing, Senate, 104th Congress, 2nd session, S. Hrg. 104-481)
AUTHOR: U.S. Congress. Senate. Committee on Governmental Affairs.
LOCATION: LRS96-4630   CONGRESSIONAL PUBLICATION   RBC 9937

SUBJECT(S): Nuclear exports--[U.S.] / Nuclear nonproliferation / International control of nuclear power / Peaceful uses of nuclear energy--[U.S.]--Treaties / Peaceful uses of nuclear energy--[Europe]--Treaties / Euratom

TITLE: Beyond CoCom: a comparative study of five national export control systems and their implications for a multilateral nonproliferation regime.
SOURCE: Comparative strategy, v. 5, Jan.-Mar. 1996: 41-57.
AUTHOR: Rudney, Robert. Anthony, T. J.
LOCATION: LRS96-5287   DK 509

NOTES: "The effectiveness of the CoCom successor regime to limit the proliferation of weapons of mass destruction will depend on the adoption of clear and comprehensive export control lists and thorough consistent implementation and enforcement by all its members. To achieve this objective, it is necessary to understand the strengths and weaknesses of the export control systems of the major supplier countries . . . . This article compares the export control systems of Germany, the United Kingdom, France, Italy, and Japan as cas studies and emphasizes the disparities among these systems that could undermine the multilateral nonproliferation regime."

SUBJECT(S): Nuclear nonproliferation / Export controls / Nuclear weapons / Arms control / North Atlantic Treaty Organization. Coordinating Committee on Multilateral Export Controls

TITLE:	Managing nuclear proliferation: condemn, strike, or assist?
SOURCE:	International studies quarterly, v. 40, 1996: 209-234.
AUTHOR:	Feaver, Peter D. Niou, Emerson, M. S.
LOCATION:	LRS96-5925 Optical Disk
NOTES:	"In order to manage proliferation, the U.S. could continue to uphold the regime, hoping to persuade the proliferator to return to non-nuclear status. It could attack, thereby ensuring that the proliferator is unable to join the nuclear club. Or it could concede the nonproliferation goal and render assistance to address the attendant safety concerns . . . . Three factors are important in determining the right option . . . . We analyze the special case of proliferation by a small enemy of the United States such as North Korea."
SUBJECT(S):	Nuclear nonproliferation--Research / Nuclear weapons / Nuclear weapons--[North Korea] / Military policy--[U.S.] / Arms control

TITLE:	Nuclear non-proliferation. Hearing, 104th Congress, 1st session. Mar. 14, 1996.
SOURCE:	Washington, G.P.O., 1996. 40 p. (Hearing, Senate, 104th Congress, 1st session, S. Hrg. 104-519)
AUTHOR:	U.S. Congress. Senate. Committee on Governmental Affairs.
LOCATION:	LRS96-5944 CONGRESSIONAL PUBLICATION RBC 10148
NOTES:	"The fate of the Nuclear Non-Proliferation Treaty . . . . Ambassador Graham is Special Representative of the President for Arms Control, and Disarmament. He will head the U.S. delegation to the 1995 Extension Conference of the Non-Proliferation Treaty."
SUBJECT(S):	Nuclear nonproliferation / Arms control agreements / Nuclear weapons / Treaty on the Non-proliferation of Nuclear Weapons

TITLE:	Europe and the challenge of proliferation.
SOURCE:	Paris, Institute for Security Studies, Western European Union, 1996. 71 p. (Chaillot paper 24)
LOCATION:	LRS96-6007 Optical Disk
NOTES:	Contents.--Introduction, by Paul Cornish, Peter Van Ham and Joachim Krause.--The proliferation of weapons of mass destruction: the risks for Europe, by Joachi Krause.--The proliferation of conventional arms and dual-use technologies, by Yves Boyer.--European nuclear non-proliferation after the NPT extension: achievements, shortcomings and needs, by Harald Muller.--Conventional arms transfers: difficulties of creating a multilateral control regime, by Christophe Carle.--European arms export

controls, by Geoffrey Van Orden.

SUBJECT(S): Nuclear nonproliferation--[Europe] / Weapons of mass destruction--[Europe] / Conventional weapons--[Europe] / Military weapons--[Europe] / Arms sales--[Europe] / Export controls--[Europe] / Arms control--[Europe]
ADDED ENTRY: Western European Union. Institute for Security Studies.

TITLE: Deep-strike weapons and strategoc stability.
SOURCE: Orbis, v. 40, fall 1996: 4989-515.
AUTHOR: Benson, Sumner.
LOCATION: LRS96-6113   Optical Disk

NOTES: "Russia and the Western powers agree that long-range conventional weapons are central to modern warfare but disagree as to whether such weapons are primarily defensive or primarly offensive. The United States and its allies have developed and deployed deep-strike weapons to counter, first, an armored assault by the Warsaw Pact against NATO and, later an atrack by nations such as Iraq and North Korea against Western friends in the Persian Gulf and East Asia."

SUBJECT(S): Arms control / Nuclear nonproliferation / Military policy--[U.S.] / Military policy--[Russia]

TITLE: Non-proliferation is embraded by Brazil.
SOURCE: Jane's intelligence review, v. 8, June 1996: 283-287.
AUTHOR: Bowen, Wyn. Koch, Andrew.
LOCATION: LRS96-6275   Optical Disk

NOTES: "Once targeted by the international non-proliferation regime for its tolerance of technology transfers to 'rogue' states like Iraq, Brazil recently completed a volte-face in its policy toward controlling the spread of unconventional weapons."

SUBJECT(S): Nuclear nonproliferation--[Brazil] / Military policy--[Brazil]

TITLE: The Treaty of Pelindaba: Africa is nuclear-weapon-free.
SOURCE: Security dialogue, v. 27, June 1996: 185-200.
AUTHOR: Ogunbanwo, Sola.
LOCATION: LRS96-6599   UV For. Africa

NOTES: "On 11 April 1996, the African heads of state and foreign miisters met in Cairo in the historic signing ceremony of the Treaty of Pelindaba .... The African nuclear-weapon-free zone (NWFZ) embraces an entire inhabited continent and will comprise 53 sovereign states."

SUBJECT(S):	Nuclear-weapon-free zones--[Africa] / Nuclear nonproliferation--[Africa] / Arms control agreements--[Africa]

TITLE:	Japan's nuclear future: the plutonium debate and East Asian security. Edited by Selig S. Harrison.
SOURCE:	Washington, Carnegie Endowment for International Peace, 1996. 12 p.
LOCATION:	LRS96-6727 Optical Disk
NOTES:	Contents.--Japan and nuclear weapons, by Seslig S. Harrison.--Why plutonium is a "must" for Japan, by Atsuyuki Suzuki.--Japan's plutonium program: a critical review, by Jinzaburo Takagi.--Japanese ambitions, U.S. constraints, and South Korea's nuclear future.
SUBJECT(S):	National security--[Japan] / Plutonium--[Japan] / Nuclear weapons--[Japan] / Nuclear nonproliferation / National security--[East Asia] / Military policy--[Japan] / Treaty on the Non-proliferation of Nuclear Weapons
ADDED ENTRY:	Harrison, Selig S.

TITLE:	Brazil and the NPT: resistance to change?
SOURCE:	Security dialogue, v. 27, Sept. 1996: 337-347.
AUTHOR:	Wrobel, Paulo S.
LOCATION:	LRS96-6764 Optical Disk
NOTES:	"In May 1995, more than 170 contracting parties decided to extend indefinitely the Treaty on the Non-Proliferation of Nuclear Weapons (NPT)."
SUBJECT(S):	Nuclear nonproliferation / Arms control agreements / Foreign relations--[Brazil] / Military policy--[Brazil]

TITLE:	Nuclear politics.
SOURCE:	International journal, v. 51, summer 1996: 397-528.
LOCATION:	LRS96-6769 LIMITED AVAILABILITY Optical Disk
NOTES:	Contents.--From winning weapon to destroyer of worlds: the nuclear taboo in international politics, by Peter Gizewski.--Rogue states and the international nuclear order, by Ashok Kapur.--Strengthening the non-proliferation regime: the role of coercive sanctions, by T. V. Paul.--The last bang before a total ban: French nuclear testing in the Pacific, by Ramesh Thakur.--Is business booming? Canada's nuclear reactor export policy, by Duane Bratt.--Going fission: tales and truths about Canada's nuclear weapons, by Don Munton.
SUBJECT(S):	Nuclear weapons / Nuclear nonproliferation / Nuclear exports--[Canada] / Nuclear weapons tests--[France] /

Nuclear weapons--[Canada]

TITLE: CTB Treaty opened for signature after approval by United Nations.
SOURCE: Arms control today, v. 26, Sept. 1996: 21-23.
AUTHOR: Cerniello, Craig.
LOCATION: LRS96-8462   Optical Disk

NOTES: "In a truly historic moment in arms control, President Clinton along with representatives of 70 other states--including the four other declared nuclear-weapon states (Britain, China, France and Russia)--signed the Comprehensive Test Ban (CTB) Treaty on September 24 at the United Nations in New York. The next day, Israel--one of the three nuclear 'threshold' states (India, Israel and Pakistan)--signed the treaty. As of September 30, 94 states have signed the treaty."

SUBJECT(S): Arms control agreements / Nuclear nonproliferation / Nuclear weapons tests

TITLE: Physics of plutonium recycling: vol. I; issues and perspectives.
SOURCE: Paris, Organisation for Economic Co-operation and Development, 1996. 190 p.
AUTHOR: OECD Nuclear Energy Agency.
LOCATION: LRS96-8761   LIMITED AVAILABILITY   L SDI Loan

NOTES: "The purpose of this Study is to review the physics of plutonium recycle as it stands today and to identify what tasks remain to be done to support future plutonium recycle strategies."

SUBJECT(S): Plutonium--[OECD countries] / Reactor fuel reprocessing--[OECD countries] / Nuclear physics--[OECD countries] / Nuclear engineering--[OECD countries] / Radioactive wastes--[OECD countries] / Nuclear nonproliferation

TITLE: Rethinking Indo-Pakistani nuclear relations: condemned to nuclear confrontation.
SOURCE: Asian survey, v. 36, June 1996: 561-583.
AUTHOR: Carranza, Mario E.
LOCATION: LRS96-8781   Optical Disk

NOTES: "This essay argues that the changes in the international system brought about by the end of the Cold War will eventually compel India and pakistan in the direction of a nuclear settlement and that they can achieve progress toward denuclearization even without a comprehensive resolution of the Kashmir dispute."

SUBJECT(S): Nuclear nonproliferation--[India] / Nuclear nonproliferation--[Pakistan]

TITLE: Swiss company investigated in Iraq A-bomb affair.
SOURCE: Middle East, no. 262, Dec. 1996: 21.
AUTHOR: George, Alan.
LOCATION: LRS96-9731 Optical Disk

NOTES: "The Vienna-based International Atomic Energy Agency (IAEA) is to ask the Swiss authorities to investigate a St Gallen company which exported uranium enrichment equipment destined for Saddam Hussein's atomic weapons programme."

SUBJECT(S): Atomic bomb / Nuclear nonproliferation / Uranium--[Iraq]

TITLE: Fighting proliferation: new concerns for the nineties. Edited by Henry Sokolski.
SOURCE: Maxwell Air Force Base, Ala., Air University Press, for sale by the Supt. of Docs., G.P.O., 1996. p. 1-129.
LOCATION: LRS96-9997 Optical DiskL CDU-D

NOTES: Partial contents.--What does the history of the nuclear nonproliferation treaty tell us about its future? by Henry Sokolski.--A nuclear nonproliferation treaty for missiles? by Richard H. Speier.--How to defeat the United States: the operational military effects of the proliferation of weapons of precise destruction, by David Blair. Part 3 is LRS96-
Part 1 of 3. Part 2 is LRS96-9998. Part 3 is LRS96-9999.

SUBJECT(S): Nuclear nonproliferation
ADDED ENTRY: Sokolski, Henry.

TITLE: Fighting proliferation: new concerns for the nineties. Edited by Henry Sokolski.
SOURCE: Maxwell Air Force Base, Ala., Air University Press, for sale by the Supt. of Docs., G.P.O., 1996. p. 130-256.
LOCATION: LRS96-9998 Optical Disk

NOTES: Partial contents.--Proliferation of land-attack cruise missiles: prospects and policy implications, by Dennis M. Gormley and K. Scott McMahon.--US commercial satellite export control policy: a debate, by Brian Dailey and Edward McGaffigan.--Resolution of the North American nuclear issue, by Walter B. Slocombe.--The nuclear deal: what the South Koreans should be concerned about, by Victor Gilinsky.--The North Korean nuclear deal and East Asian security, by Paul Wolfowitz.--Assessing the Iranian threat, by Geoffrey Kemp.
Part 2 of 3. Part 1 is LRS96-9997. Part 3 is LRS96-9999.

SUBJECT(S): Nuclear nonproliferation
ADDED ENTRY: Sokolski, Henry.

TITLE: Fighting proliferation: new concerns for the nineties. Edited by Henry Sokolski.
SOURCE: Maxwell Air Force Base, Ala., Air University Press, for sale by the Supt. of Docs., G.P.O., 1996. p. 257-377.
LOCATION: LRS96-9999   Optical Disk

NOTES: Partial contents.--Fighting proliferation with intelligence, by Henry Sokolski.--Treaty on the nonproliferation of nuclear weapons.--Agreed framework between the United States of America and the Democratic People's Republic of Korea.--Missile technology control regime: fact sheet to accompany public announcement. Part 3 of 3. Part 1 is LRS96-9997 and Part 2 is LRS96-9998.

SUBJECT(S): Nuclear nonproliferation
ADDED ENTRY: Sokolski, Henry.

TITLE: Review of the Clinton Administration nonproliferation policy. Hearing, 104th Congress, 2nd session. June 19, 1996.
SOURCE: Washington, G.P.O., 1996. 151 p.
AUTHOR: U.S. Congress. House. Committee on International Relations.
LOCATION: LRS96-10333   CONGRESSIONAL PUBLICATION   RBC 10517

SUBJECT(S): Clinton Administration / Nuclear nonproliferation / Arms control / Technology transfer

TITLE: U.S.-Israeli relations at the crossroads.
SOURCE: Israel affairs, v. 2, spring-summer 1996: whole issue (260 p.).
LOCATION: LRS96-10366   LIMITED AVAILABILITY   L SDI Loan

NOTES: Partial contents.--America's response to the New World (dis)order, by George H. Quester.--The Clinton Administration and regional security: the first two years, by Samuel F. Wells, Jr.--Israel, the United States, and the World Order crisis: fuzzy logic and conflicting principles, by Robert L. Rothstein.--US public attitudes toward Israel: a study of the attentive and issue publics, by Shibley Telhami and Jon Krosnick.--American support for Israel: history, sources, limits, by Charles Lipson.--U.S. nuclear non-proliferation policy: implications for U.S.-Israeli relations, by Shai Feldman.--Strategic aspects of U.S.-Israeli relations, by Edward N. Luttwak.

SUBJECT(S): Foreign relations--[U.S.]--Israel / Foreign relations--

ADDED ENTRY:	[Israel]--U.S. / Foreign relations--[Middle East] / Peace negotiations--[Middle East] / Strategic importance--[Middle East] / Nuclear nonproliferation--[Middle East] / International relations Sheffer, Gabriel.

TITLE:	Deterrence, weapons of mass destruction and security assurances: a European perspective.
SOURCE:	Santa Monica, Calif., Rand, 1996. 35 p.
AUTHOR:	Piper, Martyn. Tertrais, Bruno.
LOCATION:	LRS96-10420 Optical Disk
NOTES:	"In considering changes to nuclear declaratory policy, we recognize the need to address four broad sets of issues . . . deterrence of WMD usage; non-proliferation of WMD; overall deterrence of aggression against Western interests; and finally, the value of alliances in the post-Cold War world."
SUBJECT(S):	Deterrence / Weapons of mass destruction / National security--[U.S.] / First strike (Nuclear strategy) / No first use (Nuclear strategy)--[U.S.] / Military strategy--[U.S.] / Nuclear nonproliferation
ADDED ENTRY:	Rand Corporation.

TITLE:	Proliferation: threat and response.
SOURCE:	Washington, U.S. Dept. of Defense, for sale by the Supt. of Docs., G.P.O., 1996. ca. 80 p. in various pagings.
LOCATION:	LRS96-10423 Optical Disk
NOTES:	Contents.--The regional proliferation challenge.--Northeast Asia.--The Middle East and North Africa.--The former Soviet Union: Russia, Ukraine, Kazakstan, and Belarus.--South Asia.--The transnational threat: dangers from terrorism, insurgencies, civil wars, and organized crime.--Department of Defense response.--Technical annex.--Accessible technologies.
SUBJECT(S):	Nuclear nonproliferation / Strategic forces / Ballistic missile defenses--[U.S.] / Chemical weapons / Biological weapons / Military policy--[U.S.]
ADDED ENTRY:	U.S. Dept. of Defense.

TITLE:	The tritium solution.
SOURCE:	Bulletin of the atomic scientists, v. 52, July-Aug. 1996: 41-44.
AUTHOR:	Hoodbhoy, Pervez. Kalinowski, Martin.
LOCATION:	LRS96-10494 Optical Disk
NOTES:	"A simple pact could short-circuit a possible nuclear arms race between India and Pakistan."

SUBJECT(S): Tritium--[India] / Tritium--[Pakistan] / Nuclear
nonproliferation--[India] / Nuclear nonproliferation--
[Pakistan] / Nuclear weapons--[Pakistan] / Nuclear weapons-
-[India]

TITLE: Proliferation in Northeast Asia.
SOURCE: Washington, Henry L. Stimson Center, 1996. 60 p.
(Occasional paper no. 28)
AUTHOR: Mack, Andrew.
LOCATION: LRS96-11683 Optical Disk

NOTES: Partial contents.--Nuclear programs in Northeast Asia.--
Regional factors affecting nuclear proliferation.--Chemical
and biological weapons.--Responses to proliferation.

SUBJECT(S): Nuclear nonproliferation--[Northeast Asia] / Arms control--
[Northeast Asia] / Chemical weapons--[Northeast Asia] /
Biological weapons--[Northeast Asia]
ADDED ENTRY: Henry L. Stimson Center.

TITLE: Civilian nuclear programs in India and Pakistan.
SOURCE: Santa Monica, Calif., Rand, 1996. 14 p.
AUTHOR: Chow, Brian G.
LOCATION: LRS96-11737 Optical Disk

NOTES: "The current and future plans of civilian nuclear
development in India and Pakistan. Some civilian nuclear
facilities can be use to produce highly-enriched uranium
(HEU) or plutonium, the basic ingredients for making
nuclear weapons. In fact, in spite of the two countries'
active nuclear weapons programs, they both claim that their
nuclear activities are for peaceful purposes. Next, this
paper will distinguish the proliferation-resistant
activities from those that are proliferation-prone."

SUBJECT(S): Nuclear facilities--[India] / Nuclear facilities--
[Pakistan] / Nuclear nonproliferation--[South Asia]
ADDED ENTRY: Rand Corporation.

TITLE: Beyond NPT.
SOURCE: Technology review, v. 99, May-June 1996: 64-65.
AUTHOR: Steinberg, Gerald.
LOCATION: LRS96-12436

NOTES: "Policymakers must act regionally to stem the proliferation
of nuclear weapons in the post-Cold War era."

SUBJECT(S): Nuclear weapons / Arms control / Nuclear nonproliferation

TITLE:	Japan's nuclear policy: retrospect on the immediate past, perspectives on the twenty-first century.
SOURCE:	Tokyo, Institute for International Policy Studies, 1996. 31 p. (IIPS policy paper 169E, Nov. 1996)
AUTHOR:	Imai, Ryukichi.
LOCATION:	LRS96-13277 Optical Disk
NOTES:	Contents.--Nuclear weapons and Japan.--Commercial nuclear power, international safeguards.--NPT ratification, plutonium production.--Weapons-grade material issues.-- Energy supply and demand in the twentieth century.--What nuclear policy can Japan propose?
SUBJECT(S):	Nuclear weapons--[Japan] / Nuclear power--[Japan] / Plutonium--[Japan] / Nuclear nonproliferation / Energy supplies--[Japan] / Treaty on the Non-proliferation of Nuclear Weapons
ADDED ENTRY:	Institute for International Policy Studies.

TITLE:	Counterproliferation versus nonproliferation: a case for prevention versus post factum interveniton.
SOURCE:	Fletcher forum of world affairs, v. 21, winter-spring 1997: 153-171.
AUTHOR:	Turpen, Elizabeth A. Kadner, Steven P.
LOCATION:	LRS97-382 Optical Disk
NOTES:	"Prevention of "loose nukes" requires more active steps towards accounting and control of the weapons trade."
SUBJECT(S):	Nuclear nonproliferation

TITLE:	Don't ban the bomb.
SOURCE:	Economist, v. 342, Jan. 4, 1997: 15-16.
LOCATION:	LRS97-559 Optical Disk
NOTES:	Contends that the idea "of the West giving up its nuclear arms never found majority support in the democracies as long as communists elsewhere were determined to hand on to theirs. Now the cold war is over, Soviet communism presents no threat, and the disarmers have been joined by some prominent ex-generals, statesmen and thinkers . . . . As the new abolitionists see it, the risk that a nuclear weaon might go bang unexpectedly or fall into the wrong hands now outweights whatever deterrent value these weapons may still possess."
SUBJECT(S):	Nuclear nonproliferation / Atomic bomb

TITLE: Where staving off Armageddon is all in a day's work.
SOURCE: Smithsonian, v. 27, Feb. 1997: 115-128.
AUTHOR: Wolkomir, Richard. Wolkomir, Joyce.
LOCATION: LRS97-847   Optical Disk

NOTES: "At the Center for Nonproliferation Studies, scholar-activists are tracking the spread of nuclear fuels--and making a safer world."

SUBJECT(S): Peace movements / Nuclear nonproliferation

TITLE: The next Lenin: on the cusp of truly revolutionary warfare.
SOURCE: National interest, no. 47, spring 1997: 9-19.
AUTHOR: Ikle, Fred C.
LOCATION: LRS97-913   Optical Disk

NOTES: Examines the status of nuclear nonproliferation at the end of the 20th century.

SUBJECT(S): Nuclear nonproliferation

TITLE: Costing a bomb.
SOURCE: Economist, v. 342, Jan. 4, 1997: 30-31.
LOCATION: LRS97-1002   Optical Disk

NOTES: "The peace dividend vanished years ago; now the costs of defence are becoming evident once more. The first of two special articles looks at the continuing burden of nuclear weapons--both of keeping and destroying them. The second turns to the rowing threat of missile proliferation."

SUBJECT(S): Nuclear nonproliferation / Nuclear weapons

TITLE: After the CTB . . . India's intentions.
SOURCE: Bulletin of the atomic scientists, v. 53, Mar.-Apr. 1997: 49-54.
AUTHOR: Bidwai, Praful. Vanaik, Achin.
LOCATION: LRS97-1101   Optical Disk

NOTES: "India will not accept treaties that constrain its weapons options unless the nuclear powers get serious about disarmament."

SUBJECT(S): Nuclear nonproliferation / Arms control agreements

---

TITLE: The Arms race revisited.
SOURCE: Current history, v. 96, Apr. 1997: whole issue (145-192 p.).
LOCATION: LRS97-1162   Optical Disk

NOTES:	Contents.--Arms control: the unfinished agenda, by Jack Mendelsohn.--The causes of nuclear proliferation, by Scott D. Sagan.--The ballistic misile defense debate, by John Pike.--Playing politics with the chemical weapons convention, by Amy E. Smithson.--The biological weapons threat, by Jonathan B. Tucker.--The new arms race: light weapons and international security, by Michael T. Klare.--The political economy of conventional arms proliferation, by William W. Keller.--Racing toward the future: the revolution in military affairs, by Steven Metz.
SUBJECT(S):	Arms race / Ballistic missile defenses / Chemical weapons / Nuclear nonproliferation / Biological weapons / Conventional weapons / Arms control

TITLE:	A concurrent resolution expressing the sense of the Congress regarding a private visit by President Lee Teng-Hui of the Republic of China on Taiwan to the United States and a joint resolution relating to the United States-North Korea agreed framework and the obligations of North Korea under that and previous agreements with respect to the denuclearization of the Korean Peninsula and dialogue with the Republic of Korea. Markup, 104th Congress, 1st session on H.J.Res. 83 and H.Con.Res. 53. Apr. 5, 1995.
SOURCE:	Washington, G.P.O., 1997. 33 p.
AUTHOR:	U.S. Congress. House. Committee on International Relations. Subcommittee on Asia and the Pacific.
LOCATION:	LRS97-1436 AVAILABLE FROM COMMITTEE RBC 10913
SUBJECT(S):	Foreign relations--[U.S.]--Taiwan / Foreign relations--[Taiwan]--U.S. / Foreign relations--[U.S.]--North Korea / Foreign relations--[North Korea]--U.S. / Nuclear nonproliferation--[North Korea] / Foreign relations--[South Korea]--North Korea / Foreign relations--[North Korea]--South Korea / Lee, Teng-Hui.

TITLE:	Table of contents page.
SOURCE:	Bulletin of the atomic scientists, v. 53, May-June 1997.
LOCATION:	LRS97-2471 Optical Disk
NOTES:	Copy of the contents page of the journal annotated with document identification numbers (LRS numbers) to facilitate the retrieval of individual articles from the CRS optical disk system. To view the table of contents on the SCORPIO system, browse the title of the journal, select the set and display the most recent citations.
SUBJECT(S):	Nuclear nonproliferation

TITLE:	The future of nuclear power.
SOURCE:	Nuclear news, v. 40, Mar. 1997: 58-60.
AUTHOR:	Starr, Chauncey.
LOCATION:	LRS97-2697 Optical Disk
NOTES:	The magazine's editor notes ""Chauncey Starr is president emeritus of the Electric Power Research Institute, and a former ANS president (1958-59). This article is based on a speech he presented in September 1996 at a symposium at Argonne National Laboratory, observing the facility's 50th anniversary--'Research Challenges: The Next 50 Years.'"
SUBJECT(S):	Nuclear power--[U.S.]--Future / Nuclear power plants--[U.S.] / Nuclear energy policy--[U.S.] / Nuclear nonproliferation--[U.S.]

TITLE:	Continuation of the national emergency with respect to weapons of mass destruction; communication from . . . transmitting notification that the national emergency with respect to the proliferation of nuclear, biological, and chemical weapons ("Weapons of Mass Destruction"--(WMD)) and the means of delivering such weapons is to continue in effect beyond November 14, 1996--received in the United States House of Representatives November 14, 1996--received in the United States House of Representatives November 12, 1996, pursuant to 50 U.S.C. 1622(d)
SOURCE:	Washington, G.P.O., 1997. 7 p. (Document, House, 105th Congress, 1st session, no. 105-10)
AUTHOR:	U.S. President (1993- : Clinton).
LOCATION:	LRS97-4020 AVAILABLE FROM DOCUMENT ROOM Cong docs-LM220
SUBJECT(S):	Weapons of mass destruction / Nuclear nonproliferation / Arms control

TITLE:	Proliferation threats and missile defense responses. Hearing before the Military Procurement Subcommittee joint with Military Research and Development Subcommittee of the Committee on National Security, House of Representatives, 104th Congress, 1st and 2nd session. Apr. 1995-Sept. 27, 1996.
SOURCE:	Washington, G.P.O., 1997. 721 p.
AUTHOR:	U.S. Congress. House. Committee on National Security. Military Procurement Subcommittee.
LOCATION:	LRS97-4169 AVAILABLE FROM COMMITTEE RBC 11212
NOTES:	At head of title: H.N.S.C. no. 104-49. "What kinds of advances have we made . . . over the past two decades in terms of missile defense technology."
SUBJECT(S):	Ballistic missile defenses--[U.S.] / Nuclear nonproliferation / Antimissile missiles--[U.S.] / Strategic Defense Initiative

ADDED ENTRY:	U.S. Congress. House. Committee on National Security. Military Research and Development Subcommittee.

TITLE:	Scientists jointly focus on safeguarding stockpile.
SOURCE:	Aviation week and space technology, v. 146, June 23, 1997: 36-37, 40-43, 45-47, 50-51, 53-55, 57.
AUTHOR:	Scott, William B.
LOCATION:	LRS97-4856 Optical Disk
NOTES:	Contents:--Scientists jointly focus on safeguarding stockpile, by William B. Scott.--Classification sensitivities slow weapon dismantlement, by William B. Scott.--Dangers mount despite cooperative efforts, by Joseph C. Anselmo.--Superblock' typifies U.S. security measures, by William B. Scott.--Mutual trust buoys lab-to-lab exchanges, by William B. Scott.
SUBJECT(S):	International control of nuclear power / Nuclear security measures--[U.S.] / Nuclear security measures--[Russia] / Nuclear nonproliferation / Nuclear weapons--[U.S.] / Nuclear weapons--[China] / Nuclear energy research facilities--[Russia] / Nuclear energy research facilities--[U.S.] / Plutonium

TITLE:	Nuclear nonproliferation: implementation of the U.S./North Korean Agreed Framework on nuclear issues; report to the chairman, Committee on Energy and Natural Resources, U.S. Senate. June 2, 1997.
SOURCE:	Washington, G.A.O., 1997. 66 p.
AUTHOR:	U.S. General Accounting Office.
LOCATION:	LRS97-4959 Optical Disk
NOTES:	"GAO/RCED/NSIAD-97-165, B-276968" "This report discusses (1) U.S. costs to implement the Agreed Framework; (2) options for disposing of North Korea's existing spent (used) fuel; (3) the contracting for the light-water reactors and other goods and services; (4) the status of actions to normalize economic and political relations between the United States and North Korea; and (5) the status of actions to promote peace and security on the Korean Peninsula."
SUBJECT(S):	Nuclear nonproliferation / Nuclear weapons--[North Korea] / Nuclear facilities--[North Korea] / Foreign relations--[U.S.]--North Korea / Foreign relations--[North Korea]--U.S.

TITLE:	Table of contents page.
SOURCE:	Bulletin of the atomic scientists, v. 53, July-Aug. 1997.
LOCATION:	LRS97-5009 Optical Disk

NOTES:	Copy of the contents page of the journal annotated with document identification numbers (LRS numbers) to facilitate the retrieval of individual articles from the CRS optical disk system. To view the table of contents on the SCORPIO system, browse the title of the journal, select the set and display the most recent citations.
SUBJECT(S):	Nuclear nonproliferation

TITLE:	Let's reprocess the MOX plan.
SOURCE:	Bulletin of the atomic scientists, v. 53, July-Aug. 1997: 15-17.
AUTHOR:	Miller, Marvin. Von Hippel, Frank.
LOCATION:	LRS97-5010 Optical Disk
NOTES:	Criticizes the U.S. Dept. of Energy's plan to dispose of excess U.S. weapons plutonium "by fabricating it into 'mixed-oxide' (MOX) fuel, which would be burned in civilian power reactors. Critics of the MOX option say that it will undercut the long-standing U.S. opposition to the reprocessing of spent power reactor fuel and the recycling of extracted plutonium. They also argue that disposition could be accomplished more quickly and cheaply by immobilization alone."
SUBJECT(S):	Reactor fuel reprocessing--[U.S.] / Plutonium / Radioactive waste disposal--[U.S.] / Nuclear weapons--[Russia] / Nuclear weapons--[U.S.] / Nuclear nonproliferation / U.S. Dept. of Energy

TITLE:	Potential, not proliferation.
SOURCE:	Bulletin of the atomic scientists, v. 53, July-Aug. 1997: 48-53.
AUTHOR:	Mack, Andrew.
LOCATION:	LRS97-5016 Optical Disk
NOTES:	"Northeast Asia has several nuclear-capable countries, but only China has built weapons .... Japan, Taiwan, and North and South Korea all have the technical expertise to be considered virtual nuclear powers who could acquire nuclear weapons in a relatively short period of time. So far all have chosen--or have been coerced or persuaded--not to do so."
SUBJECT(S):	Nuclear weapons--[East Asia] / Arms race--[East Asia] / Nuclear nonproliferation--[East Asia]

TITLE:	Developments concerning the national emergency with respect to the proliferation of nuclear, biological, and chemical weapons ("weapons of mass destruction") and of the means of delivering such weapons; message.
SOURCE:	Washington, G.P.O., 1997. 9 p. (Document, House, 105th Congress, 1st session, no. 105-94)
AUTHOR:	U.S. President (1993- : Clinton).
LOCATION:	LRS97-5080 AVAILABLE FROM DOCUMENT ROOM Cong docs-LM220
SUBJECT(S):	Weapons of mass destruction / Nuclear nonproliferation

TITLE:	Iran's nuclear puzzle.
SOURCE:	Scientific American, v. 276, June 1997: 62-65.
AUTHOR:	Schwarzbach, David A.
LOCATION:	LRS97-5334 Optical Disk
NOTES:	"Rich in fossil-fuel resources, Iran is pursuing a nuclear power program difficult to understand in the absence of military motives."
SUBJECT(S):	Nuclear research--[Iran] / Nuclear nonproliferation--[Iran] / Nuclear power plants--[Iran] / Nuclear exports--[Russia]

TITLE:	Non-proliferation incentives for Russia and Ukraine.
SOURCE:	London, Royal United Services Institute for Defence Studies, 1997. 91 p. (Adelphi paper 309)
AUTHOR:	Baker, John C.
LOCATION:	LRS97-6124 Optical Disk
NOTES:	Contents.--Proliferation risks: formulating a new strategy. --Redirecting Ukraine's missile industries.--Improving Minatom's export policies.--Conclusion.
SUBJECT(S):	Nuclear nonproliferation--[Russia] / Nuclear nonproliferation--[Ukraine] / Export controls--[Russia] / Export controls--[Ukraine] / Nuclear exports--[Russia] / Arms control--[Russia] / Arms control--[Ukraine]

TITLE:	Integrating counterproliferation into defense planning.
SOURCE:	Santa Monica, Calif., Rand, 1997. 10 p.
AUTHOR:	Treverton, Gregory F. Bennett, Bruce W.
LOCATION:	LRS97-6153 Optical Disk
NOTES:	"The United States has long sought to halt the spread of nuclear, biological, and chemical (NBC) weapons (also called weapons of mass destruction, or WMD). It has met with substantial sucess, particularly on the nuclear front. Yet today, two developments have reshaped the counter-proliferation challenge for U.S. defense planning. The first is that many nations either have or could have NBC weapons. The second is the United States' success in

waging conventional war."
SUBJECT(S): National defense--[U.S.]--Planning / Nuclear nonproliferation / Weapons of mass destruction
ADDED ENTRY: Rand Corporation.

TITLE: Management of separated plutonium: the technical options.
SOURCE: Paris, OECD Nuclear Energy Agency, 1997. 160 p.
LOCATION: LRS97-6412   LIMITED AVAILABILITY   L SDI Loan

NOTES: "Stocks of separated plutonium in the civil nuclear fuel cycle are currently increasing. The technologies available to handle, use and dispose of it are of considerable current interest. This report presents a consensus view of experts on current and possible future technologies, based on over two decades of industrial experience of using plutonium."

SUBJECT(S): Plutonium--[OECD countries] / Nuclear nonproliferation--[OECD countries] / Reactor fuel reprocessing--[OECD countries] / Nuclear security measures--[OECD countries]
ADDED ENTRY: OECD Nuclear Energy Agency.

TITLE: Speculations on the nuclear future: possibilities, pathways, and policy implications.
SOURCE: Stanford, Calif., Asia/Pacific Research Center, 1997. 41 p. (Strategy & policy series, no. 1)
AUTHOR: Dunn, Lewis A.
LOCATION: LRS97-6470   Optical Disk

NOTES: Contents.--Today's nuclear world.--The Cold War nuclear overhang.--The proliferation dimension.--Alternative nuclear futures.--Nuclear aging.--Nuclear revivalism.--Nuclear anarchy.--A nuclear "free" world.--Some policy implications.--Which nuclear future?--Policy challenges.

SUBJECT(S): Nuclear weapons--Future / Nuclear nonproliferation--Future / Arms control--Future
ADDED ENTRY: Science Applications International Corporation. Center for Global Security and Cooperation.

TITLE: Non-proliferation: expansion of export control mechanisms.
SOURCE: Aussenpolitik, v. 48, no. 2, 1997: 137-147.
AUTHOR: Lundbo, Sten.
LOCATION: LRS97-6818   Optical Disk

NOTES: "One of the challenges to international security in the new post Cold War period is proliferation of weapons that are particularly dangerous to mankind. A crucial means to cope with this challenge is appropriate export control."

SUBJECT(S):	Nuclear nonproliferation

TITLE:	Securing the strength of the renewed NPT: China, the Linchpin "Middle Kingom".
SOURCE:	Vanderbilt journal of transnational law, v. 30, May 1997: 539-578.
AUTHOR:	Meise, Gary J.
LOCATION:	LRS97-7285 Optical Disk
NOTES:	"This Note explores China's past role in nuclear proliferation and its reasons for not acceding to the NPT regime . . . . The Note discusses inherent weaknesses of the NPT, such as its inadequate enforcement and monitoring provisions."
SUBJECT(S):	Nuclear nonproliferation--[China] / Arms control / Treaty on the Nonproliferation of Nuclear Weapons

TITLE:	Nuclear arms control.
SOURCE:	Washington quarterly, v. 20, summer 1997: 77-210.
AUTHOR:	Mazarr, Michael J.
LOCATION:	LRS97-7286 Optical Disk
NOTES:	Partial contents.--Facing nuclear reality, by Fred C. Ikle. --Is it time to junk our nukes? by Paul H. Nitze.-- Nonproliferation and U.S. nuclear policy, by James R. Schlesinger.--Russia's approach to nuclear weapons, by Nikolai Sokov.--The German debate on nuclear weapons and disarmament, by Harold Muller, and others.--The General's bombshell: phasing out the U.S. nuclear arsenal, by George Lee Butler.--The future of nuclear weapons in world affairs, by Michael Quinlan.--The abolition, by Jonathan Schell.--Thoughts about virtual nuclear arsenals, by Kenneth Waltz.--International control of nuclear weapons, by Roger D. Speed.--Avoiding nuclear anarchy, by Graham T. Allison, Owen R. Cote Jr., Richard A. Falkenrath, and Steven E. Miller.--Look before you leap: practical steps toward nuclear arms reduction, by Clifford E. Singer.
SUBJECT(S):	Arms control / Nuclear weapons--Future / Nuclear nonproliferation

TITLE:	Targets of opportunity.
SOURCE:	Bulletin of the atomic scientists, v. 53, Sept.-Oct. 1997: 22-28.
AUTHOR:	Kristensen, Hans.
LOCATION:	LRS97-7333 Optical Disk
NOTES:	Examines the issue of nuclear targeting, which includes "regional troublemakers armed with 'weapons of mass destruction.'"

SUBJECT(S): Nuclear weapons / Nuclear nonproliferation / Weapons of
mass destruction

TITLE: Deterrence and Middle East stability: an Israeli
            perspective.
SOURCE: Security dialogue, v. 28, Mar. 1997: 49-56.
AUTHOR: Steinberg, Gerald M.
LOCATION: LRS97-7407   Optical Disk

NOTES: "In the past few years, international perspectives on
            Israeli nuclear policy have begun to shift from an emphasis
            on universality in the Nuclear Non-Proliferation Treaty
            (NPT) to greater understanding of the role of this policy
            in regional deterrence and stability. Israel remains one
            of ten states that have not acceded to the NPT and one of
            three nuclear threshold states (in addition to India and
            Pakistan). However, in a number of capitals, there is more
            acceptance of the legitimacy of the nuclear ambiguity,
            given Israel's unique security requirements and the
            continuing regional threats from massive conventional
            forces, as well as chemical, biological, and nuclear
            weapons programs."

SUBJECT(S): Nuclear nonproliferation--[Middle East] / National security
            --[Israel]

TITLE: Nuclear weapons and arms control in the Middle East.
SOURCE: Cambridge, Mass., MIT Press, 1997. 336 p.
AUTHOR: Feldman, Shari.
LOCATION: LRS97-7633   LIMITED AVAILABILITY   L SDI Loan

NOTES: "This volume assesses the prospects that Middle East states
            might agree to apply some measures to arrest the spread of
            nuclear weapons. It examines the current state of nuclear
            programs in the region, the parties' perceptions of the
            danger entailed in nuclear spread, the approach of Middle
            East states to both arms control in general and to nuclear
            arms control, the implications of various items on the
            regional and global nuclear nonproliferation agenda, and
            the special role of the United States in the Middle East
            and its approach to the possible spread of nuclear weapons
            there."

SUBJECT(S): Arms control--[Middle East] / Nuclear weapons--[Middle
            East] / Nuclear nonproliferation--[Middle East]

TITLE:	Nuclear nonproliferation and safety: concerns with the International Atomic Energy Agency's technical cooperation program; report to congressional requestors. Sept. 16, 1997.
SOURCE:	Washington, G.A.O., 1997. 44 p.
AUTHOR:	U.S. General Accounting Office.
LOCATION:	LRS97-7995 Optical Disk
NOTES:	"GAO/RCED-97-192, B-277303" "While the United States and other IAEA major donor countries believe that applying safeguards is IAEA's most important function, most developing countries believe that receiving technical assistance through IAEA's technical cooperation program is just as important."
SUBJECT(S):	Nuclear nonproliferation / Arms control / Nuclear facilities--Safety measures / Technical assistance--Developing countries / International Atomic Energy Agency / Treaty on the Nonproliferation of Nuclear Weapons

TITLE:	Multilateral diplomacy at the 1995 NPT review and extension conference.
SOURCE:	Diplomacy & statecraft, v. 18, July 1997: 167-190.
AUTHOR:	Leigh-Phippard, Helen.
LOCATION:	LRS97-9035 Optical Disk
NOTES:	Discusses actions taken in 1995 to extend the Treaty on the Non-Proliferation of Nuclear Weapons (NPT).
SUBJECT(S):	Arms control agreements / Nuclear nonproliferation / Nuclear Nonproliferation Treaty (NPT)

TITLE:	Threat reduction: a framework for the future of nuclear arms control.
SOURCE:	Strategic review, v. 25, summer 1997: 46-52.
AUTHOR:	Hogler, Joe L.
LOCATION:	LRS97-9494 Optical Disk
NOTES:	"While improved relations between the U.S. and Russia make general nuclear war less likely, the threat of nuclear proliferation has grown as a real concern. Significant progress has been made in reducing and eliminating nuclear delivery systems, but the disposition of warheads and fissile material has not been addressed . . . . This article examines one roadmap to the future--a systematic framework approach to guide U.S. policy in the threat reduction arena."
SUBJECT(S):	Nuclear nonproliferation / Arms control agreements

TITLE: Thinking about nuclear weapons.
SOURCE: London, Royal United Services Institute for Defence
               Studies, 1997. 84 p. (RUSI whitehall papers 41)
AUTHOR: Quinlan, Michael.
LOCATION: LRS97-10356   Optical Disk

NOTES: Contents.--The nuclear revolution.--Deterrence and
               doctrine.--Risks, costs and their management.--After the
               Cold War.--Conclusion.

SUBJECT(S):   Nuclear weapons / Deterrence / Military policy--[NATO
               countries] / Arms control / Nuclear nonproliferation

TITLE: The May 1990 nuclear "crisis": an Indian perspective.
SOURCE: Studies in conflict and terrorism, v. 20, Oct.-Dec. 1997:
               317-332.
AUTHOR: Bhaskar, C. Uday.
LOCATION: LRS97-12106   Optical Disk

NOTES: "The events in South Asia in May 1990 are usually
               remembered as a serious crisis, a period when India and
               Pakistan could have slipped into a nuclear confrontaton but
               for the timely visit of Robert Gates, special emissary of
               then U.S. President George Bush. This nuclear-precipice
               scenario has dominated subsequent perceptions and
               interpretations of the region. However, the U.S. anxiety
               about a near-shooting nuclear war in that period stands in
               contrast to the view from India and Pakistan. Most of the
               principal participants on the subcontinent have maintained
               that they were not aware of any nuclear crisis in May
               1990."

SUBJECT(S):   Nuclear nonproliferation--[India] / Nuclear
               nonproliferation--[Pakistan]

TITLE: Nuclear arms control.
SOURCE: Washington quarterly, v. 20, summer 1997: 77-210.
LOCATION: LRS97-12130   Optical Disk

NOTES: Partial contents.--The declining utility of nuclear
               weapons, by Goodpaster Committee.--The German debate on
               nuclear weapons and disarmament, by Muller, and others.--
               The future of nuclear weapons in world affairs, by Michael
               Quinlan.--Nuclear deterrence and regional proliferation, by
               Robert G. Joseph.--International control of nuclear
               weapons, by Roger D. Speed.

SUBJECT(S):   Arms control / Nuclear weapons / Nuclear nonproliferation

TITLE:	The future of U.S. nuclea weapons policy.
SOURCE:	Arms control today, v. 27, Oct. 1997: 3-5.
AUTHOR:	Burns, William F.
LOCATION:	LRS97-12161 Optical Disk
NOTES:	Former director of the Arms Control and Disarmament Agency discusses denuclearization efforts.
SUBJECT(S):	Nuclear nonproliferation / Arms control / U.S. Arms Control and Disarmament Agency

TITLE:	Nunn-Lugar's unfinished agenda.
SOURCE:	Arms control today, v. 27, Oct. 1997: 14-22.
AUTHOR:	Ellis, Jason D. Perry, Todd.
LOCATION:	LRS97-12163 Optical Disk
NOTES:	"Nunn-Lugar has three major progam areas: destruction and dismantlement; safety, security and non-proliferation; and demilitarization and defense conversion." Examines how these initiatives are being carried out.
SUBJECT(S):	Arms control agreements / Nuclear weapons / Nuclear nonproliferation

TITLE:	Proliferation: Russian case studies. Hearing, 105th Congress, 1st session. June 5, 1997.
SOURCE:	Washington, G.P.O., 1997. 47 p. (Hearing, Senate, 105th Congress, 1st session, S. Hrg. 105-237)
AUTHOR:	U.S. Congress. Senate. Committee on Governmental Affairs. Subcommittee on International Security, Proliferation, and Federal Services.
LOCATION:	LRS97-12408 AVAILABLE FROM COMMITTEE RBC 12011
NOTES:	"Issues involving proliferation of weapons of mass destruction, particularly nuclear weapons . . . . While some of the specific Russian activities are classified, many of the details are available in the open press, and it is upon those open sources that we have relied exclusively in preparing for today's hearings."
SUBJECT(S):	Nuclear nonproliferation--[Russia] / Weapons of mass destruction--[Russia] / Arms sales--[Russia] / Technology transfer--[Russia]

TITLE:	Proliferation: threat and response.
SOURCE:	Washington, U.S. Dept. of Defense, for sale by the Supt. of Docs., G.P.O., 1997. 88 p.
LOCATION:	LRS97-12487 Optical DiskL CDU-D
NOTES:	"As the new millennium approaches, the United Stgates faces a heightened prospect that regional aggressors, third-rate

armies, terrorist cells, and even religious cults will wield disproportionate power by using--or even threatening to use--nuclear, biological, or chemical weapons against our troops in the field and our people at home . . . . Through the Department of Defense Counterproliferation Initiative, DoD contributes to government-wide efforts to prevent parties from obtaining, manufacturing, or retaining these weapons."

SUBJECT(S): Military policy--[U.S.] / Nuclear nonproliferation

TITLE: Cooperation concerning peaceful uses of nuclear energy between the United States and Brazil; message.
SOURCE: Washington, G.P.O., 1997. 64 p. (Document, House, 105th Congress, 1st session, no. 105-161)
AUTHOR: U.S. President (1993-: Clinton).
LOCATION: LRS97-12598   AVAILABLE FROM DOCUMENT ROOM   Cong docs-LM220

SUBJECT(S): Peaceful uses of nuclear energy--[U.S.]--Treaties / Peaceful uses of nuclear energy--[Brazil]--Treaties / Nuclear nonproliferation

TITLE: Dismantling the Cold War: U.S. and NIS perspectives on the Nunn-Lugar cooperative threat reduction program.
SOURCE: Cambridge, Mass., MIT Press, 1997. 426 p.
LOCATION: LRS97-13052   LIMITED AVAILABILITY   L SDI Loan

NOTES: Assesses the impact of the Nunn-Lugar Comprehensive Threat Reduction Program. Presents "A frank assessment of what U.S.-NIS [Newly Independent States] cooperation has and has not accomplished."

SUBJECT(S): Cold War / Nuclear nonproliferation
ADDED ENTRY: Shields, John M., Potter, William C.

TITLE: New strategies for the Nuclear Suppliers Group (NSG)
SOURCE: Comparative strategy, v. 16, July-Sept. 1997: 305-315.
AUTHOR: Cupitt, Richard T. Khripunov, Igor.
LOCATION: LRS97-14240   Optical Disk

NOTES: "From 1990 to 1997, membership in the Nuclear Suppliers Group (NSG) increased from 24 to 34 states. As membership in the NSG grew so rapidly, questions naturally arose about the appropriate pace and scope of this expansion. Nonetheless, important suppliers, recipients, or potential conduits of sensitive nuclear items remain outside the boundaries of NSG membership."

SUBJECT(S): Nuclear nonproliferation / Nuclear weapons

TITLE:	The contribution of the Medium Extended Air Defense System (MEADS) to U.S. post-Cold War strategy.
SOURCE:	Comparative strategy, v. 16, July-Sept. 1997: 293-304.
AUTHOR:	Rudney, Robert.
LOCATION:	LRS97-14241　　　　　　　　　　　　　　　　　　　Optical Disk
NOTES:	Discusses the development of the Medium Extended Air Defense System by the U.S., Germany and Italy as a means of countering increasing regional and proliferation threats.
SUBJECT(S):	Nuclear nonproliferation / Air defenses

TITLE:	Keystone in the arch: Ukraine in the emerging security environment of Central and Eastern Europe.
SOURCE:	Washington, Carnegie for International Peace, 1997. 145 p.
AUTHOR:	Garnett, Sherman W.
LOCATION:	LRS97-14631　　LIMITED AVAILABILITY　　　　　　　　　F
NOTES:	Contents.--Domestic sources of instability and balance.--Ukrainian-Russian relations: an overview.--The Ukrainian-Russian unfinished agenda.--The emerging security environment of Central and Eastern Europe.--Ukraine and the West: lessons of nuclear disarmament.--Shaping a 'post-nuclear' Western policy.
SUBJECT(S):	Foreign relations--[Ukraine] / National security--[Europe] / Nuclear nonproliferation--[Ukraine] / Foreign relations--[Ukraine]--Russia / Foreign relations--[Russia]--Ukraine / Central Europe / Eastern Europe

TITLE:	Tritium breakthrough brings India closer to an H-bomb arsenal.
SOURCE:	Jane's intelligence review, v. 10, Jan. 1998: 29-31.
AUTHOR:	Rethinaraj, T. S. Gopi.
LOCATION:	LRS98-191　　　　　　　　　　　　　　　　　　　　Optical Disk
NOTES:	"Nested between the nuclear capabilities of China and the nuclear aspirations of Pakistan, India would seem to be in an unenviable strategic position. As T.S. Gopi Rethinaraj reports, however, a breakthrough by Indian scientiests in the economical production of tritium may have tipped the strategic scales in New Delhi's favour."
SUBJECT(S):	Nuclear weapons--[India] / Nuclear nonproliferation--[India]

TITLE: Nuclear nonproliferation and safety: uncertainties about
              the implementation of U.S.-Russian plutonium disposition
              efforts; report to the Chairman, Committee on Foreign
              Relations, U.S. Senate. Jan. 14, 1998.
SOURCE: Washington, G.A.O., 1998. 36 p.
AUTHOR: U.S. General Accounting Office.
LOCATION: LRS98-345   Optical Disk

NOTES: "GAO/RCED-98-46, B-278690"
              "The United States and Russia have accumulated large
              stockpiles of plutonium . . . . The United States is
              implementing a long-term program to achieve the disposition
              of about 50 metric tons of excess U.S. plutonium by
              converting it into forms that would eventually be suitable
              for permanent disposal."

SUBJECT(S):   Nuclear nonproliferation / Plutonium--[U.S.] / Arms control
              / Nuclear security measures / Plutonium--[Russia] / Nuclear
              reactors

TITLE: What Saddam has in his arsenal.
SOURCE: National journal, v. 30, Jan. 31, 1998: 242-243.
AUTHOR: Kitfield, James.
LOCATION: LRS98-412   Optical Disk

NOTES: "If countries cannot be convinced to forego weapons of mass
              destruction through carrot-and-stick approach of trade
              incentives and the threat of sanctions . . . then the
              second line of defense remains detection by international
              monitors and the threat of military force."

SUBJECT(S):   Weapons of mass destruction / Arms control / Weapons of
              mass destruction--[Iraq] / Arms control verification--
              [Iraq] / Foreign relations--[U.S.]--Iraq / Foreign
              relations--[Iraq]--U.S. / Nuclear nonproliferation

TITLE: Clinton moves to permit sales of nuclear products to China.
SOURCE: Congressional Quarterly weekly report, v. 56, Feb. 7, 1998:
              328.
AUTHOR: Cassata, Donna.
LOCATION: LRS98-525   Optical Disk

NOTES: "President Clinton has set the clock ticking on an
              agreement to sell U.S. nuclear equipment to China, and
              despite a high degree of congressional discomfort,
              lawmakers are unlikely to use their power to reverse the
              deal."

SUBJECT(S):   Nuclear exports--[U.S.] / Nuclear nonproliferation--[China]

TITLE:	Proposed agreement for cooperation between the United States and Kazakhstan; message transmitting the text of a proposed agreement for cooperation between the government of the United States of America and the Republic of Kazakhstan concerning peaceful uses of nuclear energy. With accompanying annex and agreed minute, pursuant to 42 U.S.C. 2153(b). Jan. 28, 1998.
SOURCE:	Washington, G.P.O., 1998. 75 p. (Document, House, 105th Congress, 2nd session, no. 105-183)
AUTHOR:	U.S. President (1993- : Clinton).
LOCATION:	LRS98-732 AVAILABLE FROM DOCUMENT ROOM Cong. Docs.-LM-220
NOTES:	"The agreement provides a comprehensive framework for peaceful nuclear cooperation between the United States and Kazakhstan under appropriate conditions and controls reflecting our common commitment to nuclear nonproliferation goals."
SUBJECT(S):	Foreign relations--[U.S.]--Kazakhstan / Foreign relations--[Kazakhstan]--U.S. / International cooperation / Peaceful uses of nuclear energy--[Kazakhstan]--International cooperation / Nuclear nonproliferation--[Kazakhstan]

TITLE:	Proposed agreement for cooperation between the U.S. and the Swiss Federal Council; message.
SOURCE:	Washington, G.P.O., 1998. 113 p. (Document, House, 105th Congress, 2nd session, no. 105-184)
AUTHOR:	U.S. President (1993- : Clinton).
LOCATION:	LRS98-797 AVAILABLE FROM DOCUMENT ROOM Cong docs-LM220
SUBJECT(S):	Peaceful uses of nuclear energy--[U.S.]--Treaties / Peaceful uses of nuclear energy--[Switzerland]--Treaties / Nuclear nonproliferation

TITLE:	Organized crime czars corrupt Russia, prompt concerns of nuclear proliferation.
SOURCE:	National defense, v. 82, Mar. 1998: 22-23.
AUTHOR:	Williams, Robert H.
LOCATION:	LRS98-1111 Optical Disk
NOTES:	Examines the threat of criminal take-over of Russia's weapons of mass destruction. Examines the role of organized crime in weakening Russia's military.
SUBJECT(S):	Organized crime--[Russia] / Nuclear nonproliferation / Corruption in politics--[Russia]

TITLE: The Domenici challenge.
SOURCE: Bulletin of the atomic scientists, v. 54, Mar.-Apr. 1998:
            40-49.
AUTHOR: Domenici, Pete V.
LOCATION: LRS98-1270   Optical Disk

NOTES: "Republican Pete Domenici, the senior senator from New
            Mexico--the home of Los Alamos and Sandia national-
            laboratories--is attempting to breathe new life nto the
            nuclear debate. In a keynote speech to the American
            Nuclear Society last November 17, he addressed isues
            ranging from nuclear energy to food irradiation to weapons
            policy. We print his remarks on the following pages, along
            with four commentaries written for the Bulletin."

SUBJECT(S): Nuclear energy policy--[U.S.] / Nuclear weapons--[U.S.] /
            Nuclear nonproliferation--[U.S.] / Food irradiation--[U.S.]
            / Radioactive waste disposal--[U.S.]

TITLE: Faking it and making it.
SOURCE: National interest, v. 51, spring 1998: 67-80.
AUTHOR: Sokolski, Henry.
LOCATION: LRS98-1465   Optical Disk

NOTES: Examines how the U.S. deals with weapons proliferation.

SUBJECT(S): Nuclear nonproliferation / Arms control / Foreign relations
            --[U.S.]

TITLE: Focus on nonproliferation.
SOURCE: Foreign Service journal, v. 75, Feb. 1998: 16-37.
LOCATION: LRS98-1669   Optical Disk

NOTES: Contents.--Almost a success story, by Lawrence Scheinman.--
            The South Asian standoff, by Raju G. C. Thomas.--The
            Iran/Iraq conundrum, by George Gedda.

SUBJECT(S): Nuclear nonproliferation / Nuclear weapons / Arms control /
            India / Pakistan / Iraq / Iran

TITLE: Prospects for light-water nuclear reactor project.
SOURCE: Korea focus, v. 6, Jan.-Feb. 1998: 81-88.
AUTHOR: Duk-min, Yun.
LOCATION: LRS98-1867   Optical Disk

NOTES: "The signing of the Agreed Framework between the United
            States and North Korea on October 21, 1994, in Geneva was
            the result of 19 months of bilateral talks on the North
            Korean nuclear question precipitated by Pyongyang's avowed
            plan to withdraw from the Nuclear Nonproliferation Treaty
            (NPT)."

SUBJECT(S):	Nuclear reactors--[North Korea] / Nuclear nonproliferation--[North Korea] / Nuclear Nonproliferation Treaty (NPT)

TITLE:	Compilation of hearings on national security issues.
SOURCE:	Washington, G.P.O., 1998. 675 p. (Print, Senate, 105th Congress, 2d session, S. Prt. 105-50)
AUTHOR:	U.S. Congress. Senate. Committee on Governmental Affairs. Subcommittee on International Security, Proliferation, and Federal Services.
LOCATION:	LRS98-3255 AVAILABLE FROM COMMITTEE RBC 12588
NOTES:	Hearings, 105th Congress, 1st session. Feb. 12-Oct. 27, 1997. Contents.--The future of nuclear deterrence (Feb. 12, 1997).--National missile defense and prospects for U.S.-Russia ABM Treaty accommodation (Mar. 13, 1997).--Proliferation: Chinese case studies (Apr. 10, 1997).--National missile defense and the ABM Treaty (May 1, 1997).--Proliferation: Russian case studies (June 5, 1997.--Proliferation and U.S. export controls (June 11, 1997)..--Compliance review process and missile defense (July 21, 1997).--Missile proliferation in the information age (Sept. 22, 1997).--North Korean missile proliferation (Oct. 21, 1997).--Safety and reliability of the U.S. nuclear deterrent (Oct. 27, 1997).
SUBJECT(S):	Deterrence--Future / Strategic forces--[U.S.] / Nuclear nonproliferation--[China] / Nuclear nonproliferation--[Russia] / Nuclear exports--[U.S.] / Arms control agreements--[U.S.]--Russia / Arms control agreements--[Russia]--U.S. / Ballistic missile defenses--Treaties / Ballistic missiles--[North Korea] / Arms sales--[North Korea] / Nuclear weapons information--Technological innovations / Nuclear weapons tests--[U.S.] / ABM Treaty / Strategic Arms Reduction Treaty (START)

TITLE:	The CTB treaty and nuclear non-proliferation: the debate continues.
SOURCE:	Arms control today, v. 28, Mar. 1998: 7-11.
AUTHOR:	Keeny, Spurgeon M., Jr. Bailey, Kathleen C.
LOCATION:	LRS98-3501 Optical Disk
NOTES:	Presents edited versions of statements by Keeny and Bailey to the Senate Subcommittee on International Security, Proliferation and Federal Services hearing on March 18, 1998.
SUBJECT(S):	Arms control / Nuclear nonproliferation / Comprehensive Test Ban Treaty

TITLE: Proliferation threats through the year 2000. Hearing,
             105th Congress, 1st session. Oct. 8, 1997.
SOURCE: Washington, G.P.O., 1998. 91 p. (Hearing, Senate, 105th
             Congress, 1st session, S. Hrg. 105-359)
AUTHOR: U.S. Congress. Senate. Committee on Foreign Relations.
LOCATION: LRS98-3770   AVAILABLE FROM COMMITTEE   RBC 12604

NOTES: "Proliferation is one of the most important threats which
             face our Nation today. The danger of rogue nations
             acquiring and using, at least as leverage, a nuclear
             capability is real."

SUBJECT(S): Nuclear nonproliferation / Nuclear weapons / Arms control /
             National security--[U.S.] / Iran

TITLE: Table of contents page.
SOURCE: Bulletin of the atomic scientists, v. 54, May-June 1998.
LOCATION: LRS98-3888   Optical Disk

NOTES: Copy of the contents page of the journal annotated with
             document identification numbers (LRS numbers) to facilitate
             the retrieval of individual articles from the CRS optical
             disk system. To view the table of contents on the SCORPIO
             system, browse the title of the journal, select the set and
             display the most recent citations.

SUBJECT(S): Nuclear nonproliferation

TITLE: U.S. intelligence takes the heat for dim insight on nuclear
             proliferation.
SOURCE: CQ weekly, v. 56, June 6, 1998: 1542-1544.
AUTHOR: Pomper, Miles A.
LOCATION: LRS98-4204   Optical Disk

NOTES: "Feeble effect of economic sanctions on India and Pakistan
             has administration softening its position, not pressuring
             allies."

SUBJECT(S): Nuclear nonproliferation--[India] / Nuclear weapons tests--
             [India] / Nuclear weapons tests--[Pakistan] / Intelligence
             services--[U.S.]

TITLE: Nonproliferation approaches in the caucasus.
SOURCE: Nonproliferation review, winter 1998: 108-120.
AUTHOR: Robinson, Tamara C.
LOCATION: LRS98-4211   Optical Disk

NOTES: "This report analyzes nonproliferation approaches in
             Georgia, Armenia, and Azerbaijan by examining the following
             indicators: 1) the quantity of nuclear materials in the
             region; 2) the strength of domestic export controls; 3) the

	degree of participation in the international nonproliferation regime; and 4) the feasibility of creating a Caucasian nuclear-weapon-free zone."
SUBJECT(S):	Nuclear nonproliferation--[transcaucasia] / Caucasus

TITLE:	Israel and the evolution of U.S. nonproliferation policy: the critical decade (1958-1968).
SOURCE:	Nonproliferation review, winter 1998: 1-19.
AUTHOR:	Cohen, Avner.
LOCATION:	LRS98-4214 Optical Disk
NOTES:	"Israel was a powerful testimony to the Eisenhower administration's failure to come to grips with the reality of nuclear proliferation. Although the Israeli nuclear project had been conceived in 1955-57 and its physical construction initiated in early 1958, only in December 1960 did the departing Eisenhower administration determine that Israel was in fact building a major nuclear facility in the Negev desert aimed at establishing a nuclear weapons capability."
SUBJECT(S):	Nuclear nonproliferation--[Israel] / Nuclear weapons--[Israel]

TITLE:	Norms and nuclear proliferation: Sweden's.
SOURCE:	Nonproliferation review, winter 1998: 32-43.
AUTHOR:	Arnett, Eric.
LOCATION:	LRS98-4336 Optical Disk
NOTES:	Assessing how nations have dealt with Iran, this article examines the implementation of nonproliferation planning.
SUBJECT(S):	Nuclear nonproliferation--[Iran] / Arms control agreements

TITLE:	Agreement for the nuclear cooperation between the United States and China; communication. Feb. 3, 1998.
SOURCE:	Washington, G.P.O., 1998. 37 p. (Document, House, 105th Congress, 2nd session, no. 105-197)
AUTHOR:	U.S. President (1993- : Clinton).
LOCATION:	LRS98-4480 AVAILABLE FROM DOCUMENT ROOM Cong docs-LM220
SUBJECT(S):	Peaceful uses of nuclear energy--[U.S.]--Treaties / Peaceful uses of nuclear energy--[China]--Treaties / Nuclear nonproliferation--[U.S.]--Treaties / Nuclear nonproliferation--[China]--Treaties

TITLE:	Table of contents page.
SOURCE:	Bulletin of the atomic scientists, v. 54, July-Aug. 1998.
LOCATION:	LRS98-5066 Optical Disk

NOTES: Copy of the contents page of the journal annotated with
document identification numbers (LRS numbers) to facilitate
the retrieval of individual articles from the CRS optical
disk system. To view the table of contents on the SCORPIO
system, browse the title of the journal, select the set and
display the most recent citations.

SUBJECT(S): Nuclear nonproliferation / Nuclear weapons

TITLE: Nuclear nonproliferation: uncertainties with implementing
IAEA's strengthened safeguards system; report to the
chairman, Committee on International Relations, House of
Representatives. July 9, 1998.
SOURCE: Washington, G.A.O., 1998. 40 p.
AUTHOR: U.S. General Accounting Office.
LOCATION: LRS98-5246   Optical Disk

NOTES: "GAO/NSIAD/RCED-98-184, B-280004"
"International Atomic Energy Agency (IAEA) safeguards are a
cornerstone of U.S. and international efforts to prevent
nuclear weapons proliferation . . . . IAEA regularly
inspects all facilities or locations containing declared
material to verify its peaceful uses. The discovery that
Iraq had developed a clandestine nuclear weapons program
while IAEA was inspecting Iraq's civillian nuclear
facilities caused the Agency and its its member states to
initiate an intensive effort to strengthen further the
safeguards system . . . . This report describes the
changes IAEA is undertaking to strengthen its safeguards
program."

SUBJECT(S): Nuclear nonproliferation--Safety measures / Arms control /
International Atomic Energy Agency / Treaty on the
Nonproliferation of Nuclear Weapons

TITLE: Arms control in the North Pacific: the role for confidence
building and verification; Third Annual Cooperative
Research Workshop 25-27 February, 1994.
SOURCE: Victoria, British Columbia, Non-Proliferation, Arms Control
and Disarmament Division, Dept. of Foreign Affairs and
International Trade, 1994.   278 p.
LOCATION: LRS94-9853   LIMITED AVAILABILITY

NOTES: Partial contents.--Arms control in the North Pacific: the
role for confidence building and verification--third annual
workshop overview, by James Macintosh.--The emerging
security balance in the North Pacific and the nuclear
impasse on the Korean Peninsula, by James A. Boutilier.--
Resolving the North Korean nuclear issue: a South Korean
perspective, by Man -Kwon Nam.--North Korea's nuclear and
ballistic missile programs from a non-proliferation

perspective: challenge to verification, by Jin-Pyo Yoon.--
The politics of multilateral satellite reconnaissance: a
reappraisal.

SUBJECT(S): Nuclear weapons--[North Korea] / Arms control--[North
Korea] / Nuclear nonproliferation / Arms control
verification--[Northeast Asia] / Canada. Dept. of Foreign
Affairs and International Trade / Non-Proliferation, Arms
Control and Disarmament Division / North Pacific
ADDED ENTRY: Boutilier, James A.

TITLE:	North Korean problem tests Japan's evolving diplomacy.
SOURCE:	Washington, Japan Economic Institute, 1994. 14 p. (JEI report no. 24A)
AUTHOR:	Wanner, Barbara.
LOCATION:	LRS94-6686 JX 1428 For. Japan D 1

NOTES: "The crisis caused by North Korea's stubborn refusal to
comply fully with International Atomic Energy Agency
inspections under the nuclear Nonproliferation Treaty now
challenges Japan's foreign policy at an important juncture
in its postwar evolution . . . . Tokyo has worked closely
with Washington and Seoul . . . . The Japanese government
also has sought through shuttle diplomacy to enlist the
help of the People's Republic of China in resolving this
crisis."

SUBJECT(S): Foreign relations--[Japan]--North Korea / Foreign relations
--[North Korea]--Japan / Nuclear weapons--[North Korea] /
Arms control verification--[North Korea] / Nuclear
nonproliferation

TITLE:	Nuclear verification under the NPT: what should it cover-- how far may it go?
SOURCE:	Southhampton, England, Mountbatten Centre for International Studies, University of Southhampton, 1994. 26 p. (Programme for Promoting Nuclear Non-Proliferation. PPNN study no. 5)
AUTHOR:	Bunn, George. Timerbaev, Roland M.
LOCATION:	LRS94-6690 UA 17

NOTES: Partial contents.--NPT prohibitions on weaponization.--
Inspection for weaponization.--UN Security Council
Enforcement of Inspections.

SUBJECT(S): Arms control verification / Nuclear nonproliferation--
Treaties / Arms control agreements / Programme for
Promoting Nuclear Non-Proliferation

TITLE: Non-proliferation and multilateral verification: the Comprehensive Nuclear Test Ban Treaty (CTBT); symposium proceedings.
SOURCE: Toronto, Canada, Center for International and Strategic Studies, York University, 1994. 261 p.
LOCATION: LRS94-10649   LIMITED AVAILABILITY

NOTES: Partial contents.--Non-proliferation and the CTBT: the negotiating challenge.--Developing a functional CTBT verification package.--Extending the agenda.

SUBJECT(S): Nuclear weapons tests / Arms control agreements / Arms control verification / Nuclear nonproliferation / Comprehensive Nuclear Test Ban Treaty (CTBT)
ADDED ENTRY: Mataija, Steven.

TITLE: Verifying a fissile materials cut-off: an exploratory analysis of potential diversion scenarios.
SOURCE: [Canada] Dept. of Foreign Affairs and International Trade, 1994. 58 p.
LOCATION: LRS94-14969   UA 17

NOTES: "Exploration of potential diversion paths relevant to a fissile materials 'cut-off' agreement and the implications of these paths for verification."

SUBJECT(S): Fissionable materials--Scenarios / Nuclear nonproliferation / Arms control / Nuclear weapons / Arms control verification / Nuclear facilities--Scenarios
ADDED ENTRY: Winfield, David J., Campbell, Robert H.

TITLE: The nuclear deal with North Korea: is the glass half empty or half full?
SOURCE: Comparative strategy, v. 14, Apr.-June 1995: 137-148.
AUTHOR: Bailey, Kathleen C.
LOCATION: LRS95-7630   DS 930 E

NOTES: "This article traces the development of the North Korean nuclear program and the negotiations with the International Atomic Energy Agency and the United States . . . . North Korea poses a real military threat to East Asia, and its clear violations and flaunting of the Nuclear Nonproliferation Treaty endanger the nonproliferation regime."

SUBJECT(S): Nuclear weapons--[North Korea] / Nuclear research--[North Korea] / Nuclear nonproliferation / Arms control

Nuclear Proliferation: An Annotated Biography 155

TITLE: Going just a little nuclear: nonproliferation lessons from North Korea.
SOURCE: International security, v. 20, fall 1995: 92-122.
AUTHOR: Mazarr, Michael J.
LOCATION: LRS95-12727 DS 930 E

NOTES: "A brief history of the North Korean nuclear program and the U.S.-led effort to stop it . . . the lessons of the North Korean case for future nonproliferation efforts . . . and . . . an assessment of the October 1994 U.S.-North Korean Agreed Framework in the context of these lessons."

SUBJECT(S): Nuclear weapons--[North Korea] / Nuclear facilities--[North Korea] / Nuclear research--[North Korea] / Nuclear nonproliferation

TITLE: The threat of nuclear war on the Korean Peninsula.
SOURCE: Jane's intelligence review, v. 7, Sept. 1995: 418-419.
AUTHOR: Young, Peter Lewis.
LOCATION: LRS95-8107 DS 930 E

NOTES: "An attempt to end the impasse has suggested that the present North Korean facilities be progressively closed down and the DPRK be given access to modern nuclear technology. In exchange, the North would allow inspection . . . . Despite the impression of progress gained from the negotiations, any agreement may founder on internal North Korean politics."

SUBJECT(S): Nuclear facilities--[North Korea] / Nuclear weapons--[North Korea] / Nuclear research--[North Korea] / Arms control / Nuclear nonproliferation / Arms control verification--[North Korea] / Politics and government--[North Korea] / Foreign relations--[U.S.]--North Korea / Foreign relations--[North Korea]--U.S.

TITLE: North Korea Nuclear Agreement. Hearings, 104 Congress, 1st session. Jan. 24 and 25, 1995.
SOURCE: Washington, G.P.O., 1995. 119 p. (Hearing, Senate, 104th Congress, 1st session, S. Hrg. 104-125)
AUTHOR: U.S. Congress. Senate. Committee on Foreign Relations.
LOCATION: LRS95-8235 CONGRESSIONAL PUBLICATION

NOTES: "The purpose of this morning's hearing is to provide an opportunity for the administration to clarify to the American people and to Congress why it is that the agreed framework with North Korea is in the national interest of the United States."

SUBJECT(S): Nuclear weapons--[North Korea] / Nuclear facilities--[North Korea] / Nuclear nonproliferation / Arms control / Arms control verification--[North Korea] / Foreign relations--

[U.S.]--North Korea / Foreign relations--[North Korea]--
U.S.

TITLE:	Omnibus Export Administration Act of 1996; report together with additional views to accompany H.R. 361, including cost estimate of the Congressional Budget Office.
SOURCE:	Washington, G.P.O., 1996. 174 p. (Report, Senate, 104th Congress, 2nd session, no. 104-605)
AUTHOR:	U.S. Congress. Senate. Committee on International Relations.
LOCATION:	LRS96-3805 CONGRESSIONAL PUBLICATION L CDU
SUBJECT(S):	Export control--[U.S.]--Law and legislation / Nuclear nonproliferation--Treaties / Terrorism--Treaties / Omnibus Export Administration Act Proposed) / Trading with the Enemy Act

TITLE:	Global proliferation of weapons of mass destruction. Hearings, 104th Congress, 1st session. Part 1. Oct. 31-Nov. 1, 1995.
SOURCE:	Washington, G.P.O., 1996. 730 p. (Hearing, Senate, 104th Congress, 1st session, S. Hrg. 104-422, part 1)
AUTHOR:	U.S. Congress. Senate. Committee on Governmental Affairs. Permanent Subcommittee on Investigations.
LOCATION:	LRS96-3462 CONGRESSIONAL PUBLICATION RBC 9819
SUBJECT(S):	Weapons of mass destruction / Nuclear nonproliferation / Arms control / Terrorism / Aum Shinrikyo

TITLE:	Developments concerning the national emergency with respect to the proliferation of nuclear, biological, and chemical weapons ("weapons of mass destruction") and of the means of delivering such weapons; message from the President of the United States transmitting a report on the national emergency declared by Executive Order no 12938 of November 14, 1994, in response to the threat posed by the proliferation of nuclear, biological, and chemical weapons ('weapons of mass destruction') and of the means of delivering such weapons, pursuant to 50 U.S.C. 1703(e) and 50 U.S.C. 1641(c).
SOURCE:	Washington, G.P.O., 1996. 7 p. (Document, House, 104th Congress, 2d session, 104-210)
AUTHOR:	U.S. President (1993- : Clinton).
LOCATION:	LRS96-2728 CONGRESSIONAL PUBLICATION UV Gen.
SUBJECT(S):	Weapons of mass destruction / Nuclear nonproliferation / Arms control / Ballistic missiles

Nuclear Proliferation: An Annotated Biography 157

TITLE:	Five minutes past midnight: the clear and present danger of nuclear weapons grade fissile materials.
SOURCE:	Colorado Springs, U.S. Air Force, Institute for National Security Studies, 1996. 70 p. (INSS occasional paper 8; Proliferation series)
AUTHOR:	Roberts, Guy B.
LOCATION:	LRS96-2206 UC 650 A
NOTES:	Partial contents.--What is the problem?--The growing proliferation risk of fissile materials.--The non-proliferation regime as a framework for controlling fissile materials.
SUBJECT(S):	Fissionable materials / Nuclear nonproliferation / Export controls / Arms control

TITLE:	Nuclear proliferation: the diplomatic role of non-weaponized programs.
SOURCE:	Colorado Springs, U.S. Air Force Academy, Institute for National Security Studies, 1996. 30 p. (INSS occasional paper 7)
AUTHOR:	Reynolds, Rosalind R.
LOCATION:	LRS96-1553 QC 170 Gen.
NOTES:	"This paper presents a nontraditional, almost revisionist approach to the vital topic of nuclear proliferation . . . . Reynolds points out that states can achieve many of their national security goals through the mere capability of producing nuclear weapons. Existential deterrence may occur without even having any weapons, as long as the potential adversary believes that a state could develop them."
SUBJECT(S):	Nuclear nonproliferation / Nuclear facilities / Nuclear research / Nuclear engineering

TITLE:	The Nunn-Lugar Act: a wasteful and dangerous illusion.
SOURCE:	Washington, Cato Institute, 1996. 15 p. (Foreign policy briefing no. 39)
AUTHOR:	Kelley, Rich.
LOCATION:	LRS96-1505 JX 1435 U.S. CIS countries
NOTES:	"The Cooperative Threat Reduction (CTR) program to provide assistance for dismantling or safely storing the weapons in the Soviet nuclear arsenal."
SUBJECT(S):	American military assistance--[CIS countries] / Nuclear weapons--[CIS countries] / Nuclear nonproliferation / Arms control

TITLE: On compliance with nuclear non-proliferation obligations.
SOURCE: Security dialogue, v. 27, Mar. 1996: 17-26.
AUTHOR: Elbaradei, Mohamed.
LOCATION: LRS96-1265   UA 17

NOTES: "While states are ready to assume international
              obligations, they are rarely ready to accept independent
              monitoring of their compliance with these obligations."

SUBJECT(S):   Nuclear nonproliferation / Arms control / International
              obligations / Arms control verification

TITLE: Security implications of the nuclear non-proliferation
              agreement with North Korea. Hearing, 104th Congress, 1st
              session. Jan. 26, 1995.
SOURCE: Washington, G.P.O., 1996. 95 p. (Hearing, Senate, 104th
              Congress, 1st session, S. Hrg. 104-188)
AUTHOR: U.S. Congress. Senate. Committee on Armed Services.
LOCATION: LRS96-805   CONGRESSIONAL PUBLICATION   RBC 9384

SUBJECT(S):   Nuclear nonproliferation / Nuclear weapons--[North Korea] /
              Nuclear facilities--[North Korea] / Foreign relations--
              [U.S.]--North Korea / Foreign relations--[North Korea]--
              U.S. / National security--[U.S.]

TITLE: Reports of Chinese shipments put Clinton on the spot.
SOURCE: Congressional Quarterly weekly report, v. 54, Feb. 17,
              1996: 396-397.
AUTHOR: Doherty, Carroll J.
LOCATION: LRS96-654   JX 1428 For. China B

NOTES: "Administration is unwilling to tolerate nuclear exports,
              but sanctions on Beijing could harm U.S. businesses."

SUBJECT(S):   Sanctions (International law)--[China] / Export controls--
              [China] / Nuclear nonproliferation--[China]

TITLE: Joseph Rotblat: the road less traveled.
SOURCE: Bulletin of the atomic scientists, v. 52, Jan.-Feb. 1996:
              46-54.
AUTHOR: Landau, Susan.
LOCATION: LRS96-150   CT 100 Rotblat, Joseph

NOTES: Profiles Joseph Rotblat and the Pugwash Conferences on
              Science and World Affairs, and the movement to rid the
              world of nuclear weapons.

SUBJECT(S):   Scientists / Arms control / Nuclear nonproliferation /
              Nuclear weapons / Antinuclear weapons movement / Rotblat,
              Joseph / Pugwash Conferences

TITLE:	Table of contents page.
SOURCE:	Bulletin of the atomic scientists, v. 52, Jan.-Feb. 1996.
LOCATION:	LRS96-148 Optical Disk
NOTES:	Copy of the contents page of the journal annotated with document identification numbers (LRS numbers) to facilitate the retrieval of individual articles from the CRS optical disk system. To view the table of contents on the SCORPIO system, browse the title of the journal, select the set and display the most recent citations.
SUBJECT(S):	Nuclear nonproliferation

TITLE:	Nuclear illusions: Argentina and Brazil.
SOURCE:	Washington, Henry L. Stimson Center, 1995. 49 p. (Occasional paper 25)
AUTHOR:	Redick, John R.
LOCATION:	LRS95-13398 UV For. Latin America
NOTES:	"Latin America's two leading nations have devoted considerable resources to nuclear development. Both have achieved significant progress . . . . Until recently, their nuclear development has been accompanied . . . in rejection of the basic tenets of the non-proliferation regime . . . . These policies have now been reversed . . . and both nations have apparently embraced the non-proliferation regime. The reversal in nuclear policies resulted from domestic political change and the evolution of Argentine-Brazilian relations."
SUBJECT(S):	Nuclear research--[Brazil] / Nuclear research--[Argentina] / Nuclear nonproliferation--[Latin America] / Arms control / Treaty of Tlatelolco

TITLE:	Extending the nuclear non-proliferation treaty.
SOURCE:	Behind the headlines, v. 52, spring 1995: whole issue (16 p.)
AUTHOR:	Rauf, Tariq.
LOCATION:	LRS95-13393 UA 17
NOTES:	"At the centre of postwar non-proliferation efforts is the landmark Treaty on the Non-Proliferation of Nuclear Weapons (NPT), signed in July 1968 and entered into force in 1970 . . . . Through it more than 160 sovereign non-nuclear weapons states have renounced their right to develop nuclear arms."
SUBJECT(S):	Nuclear nonproliferation--Treaties / Arms control agreements / Nuclear weapons / Treaty on Non-Proliferation of Nuclear Weapons (NPT)

TITLE: Nuclear proliferation: diminishing threat?
SOURCE: Colorado Springs, Colo., USAF Institute for National Security Studies, U.S. Air Force Academy, 1995. 56 p. (INSS occasional paper 6)
AUTHOR: Kincade, William H.
LOCATION: LRS95-13158 UC 650 A

NOTES: Partial contents.--Evolution of proliferation.--The Nth country problem revisited.--The nuclear club: technical barriers to entry.--The nuclear club: growing disincentives to join.- -Proliferation implications.--Future non-proliferation policy.

SUBJECT(S): Nuclear nonproliferation / Nuclear weapons / Arms control

---

TITLE: Disarmament: ending reliance on nuclear and conventional arms.
SOURCE: New York, United Nations, c1995. 155 p.
AUTHOR: United Nations. Centre for Disarmament affairs.
LOCATION: LRS95-13154 LIMITED AVAILABILITY L SDI Loan

NOTES: Contents.--Moving towards real nuclear disarmament.--The comprehensive test ban and the review and extension conference of the non-proliferation treaty.--The oversight capabilities of the International Atomic Energy Agency and the non-proliferation treaty.--Military expenditures and social development: missed opportunities? Future possibilities?--The UN arms register.--The land mine crisis: humanitarian disaster: what can be done?

SUBJECT(S): Arms control / Nuclear nonproliferation / Conventional weapons / Nuclear weapons tests / Land mines / U.N. Arms Register

---

TITLE: Verifying nonproliferation treaties: obligation, process, and sovereignty.
SOURCE: Washington, National Defense University Press, 1995. 155 p.
AUTHOR: Kessler, J. Christian.
LOCATION: LRS95-13051 Optical Disk

NOTES: Partial contents.--What is verification?--Controlling nuclear weapons.--Biological weapons.--Chemical weapons.--Limiting conventional forces in Europe.--Looking for patterns.

SUBJECT(S): Arms control agreements / Nuclear weapons / Biological weapons / Arms control verification / Chemical weapons / Conventional weapons / Nuclear nonproliferation / Arms control

TITLE:	KEDO takes first steps to fulfill nuclear accord with North Korea.
SOURCE:	Arms control today, v. 25, Sept. 1995: 29.
AUTHOR:	Medeiros, Evan S.
LOCATION:	LRS95-12944 DS 930 E
NOTES:	"The United States and its Asian allies made encouraging progress during August implementing the denuclearization agreement with North Korea, expanding the multilateral effort to construct two proliferation-resistant nuclear reactors in the North and conducting a politically sensitive site survey with a team that included South Korean representatives."
SUBJECT(S):	Nuclear nonproliferation--[North Korea] / Collective security--[North Korea]

TITLE:	The proposed fissile-material production cutoff: next steps.
SOURCE:	Santa Monica, Calif., Rand Corporation, 1995. 51 p. (MR-586-1-OSD)
AUTHOR:	Chow, Brian G. Speier, Richard H. Jones, Gregory S.
LOCATION:	LRS95-12544 UC 650 A
NOTES:	Partial contents.--The U.S. proposal.--Third world inventories and ability to produce weapon-usable material.--The proposed convention's effect on proliferation.--Next steps: options, obstacles, and mitigation measures.
SUBJECT(S):	Nuclear nonproliferation / Fissionable materials / Arms control
ADDED ENTRY:	Rand Corporation.

TITLE:	Space technology as a factor of international stabilization and destabilization.
SOURCE:	Space policy, v. 11, Nov. 1995: 233-238.
AUTHOR:	Becher, Klaus.
LOCATION:	LRS95-10966 QB 1 M 1
NOTES:	"From the viewpoint of the leading space powers, increased space cooperation with newcomers can actually offer an opportunity, under certain specified conditions, to prevent confrontational and destabilizing space technology proliferation. It is therefore advisable that space technology suppliers reinforce their willingness to lend active support to the peaceful goals of other countries' space programmes."
SUBJECT(S):	Space sciences / International relations / Technology and social problems / Nuclear nonproliferation / International cooperation in astronautics

TITLE:	A new nuclear threat?
SOURCE:	Moscow times, v. 2, Nov. 5, 1995: 21.
AUTHOR:	Garnett, Sherman.
LOCATION:	LRS95-10513 UA 17
NOTES:	"The pressing question for Russian nuclear doctrine is how to deter Saddam Hussein and the Saddam Husseins to come."
SUBJECT(S):	Nuclear nonproliferation / Nuclear weapons / Deterrence / Arms control

TITLE:	Nuclear successor states of the Soviet Union: nuclear weapon and sensitive export status report.
SOURCE:	Washington, Carnegie Endowment for International Peace, 1995. 79 p.
LOCATION:	LRS95-10469 F
NOTES:	Partial contents.--Nuclear status.--Export controls and sensitive exports.
SUBJECT(S):	Nuclear weapons--[CIS countries] / Nuclear weapons--[Russia] / Nuclear nonproliferation / Nuclear exports / Export controls--[CIS countries] / Arms control
ADDED ENTRY:	Carnegie Endowment for International Peace., Monterey Institute of International Studies.

TITLE:	Agreement for cooperation between the United States of America and the Republic of South Africa concerning peaceful uses of nuclear energy; message.
SOURCE:	Washington, G.P.O., 1995. 67 p. (Document, House, 104th Congress, 1st session, no. 104-121)
AUTHOR:	U.S. President (1993- : Clinton)
LOCATION:	LRS95-10358 CONGRESSIONAL PUBLICATION QC 170 Gen.
SUBJECT(S):	Peaceful uses of nuclear energy--[U.S.]--Treaties / Peaceful uses of nuclear energy--[South Africa]--Treaties / Nuclear nonproliferation

TITLE:	Dealing with North Korea: the case of the Agreed Framework.
SOURCE:	Journal of Northeast Asian studies, v. 14, summer 1995: 91-101.
AUTHOR:	Lee, Dong-bok.
LOCATION:	LRS95-10262 DS 930 E
NOTES:	"The 'Agreed Framework,' a deal that the Untied States and the DPRK cut in Geneva in October 1994 on the North Korean nuclear issue, now approaches the first of its check points to pass a test as to whether it really has a chance to survive . . . . From the many 'ambiguities' and omissions'

of the Agreed Framework, the United States now enters a
stage where it will have to brace for another wave of
North Korea's 'diplomatic brinkmanship.'"

SUBJECT(S): Nuclear weapons--[North Korea] / Nuclear research--[North
Korea] / Nuclear facilities--[North Korea] / Arms control
verification--[North Korea] / Nuclear nonproliferation /
Foreign relations--[U.S.]--North Korea / Foreign relations-
-[North Korea]--U.S.

TITLE:	The U.S.-DPRK nuclear deal: status and prospects.
SOURCE:	Korea & world affairs, v. 19, fall 1995: 482-509.
AUTHOR:	Mazarr, Michael J.
LOCATION:	LRS95-10260 DS 930 E

NOTES: "This essay . . . begins with a review of U.S. and ROK
nonproliferation policy . . . . The essay then describes
and assesses the Agreed Framework itself. It concludes
with an examination of several critical issues that remain
to be resolved."

SUBJECT(S): Nuclear weapons--[North Korea] / Nuclear facilities--[North
Korea] / Arms control / Nuclear nonproliferation / Arms
control verification--[North Korea] / Foreign relations--
[U.S.]--North Korea / Foreign relations--[North Korea]--
U.S.

TITLE:	China's nonproliferation and export control policies: boom or bust for the NPT regime?
SOURCE:	Asian survey, v. 35, June 1995: 587-603.
AUTHOR:	Davis, Zachary S.
LOCATION:	LRS95-10102 UV For. China

NOTES: "This article examine the gap between China's declared
nonproliferation policy and its failure to put an end to
controversial exports."

SUBJECT(S): Nuclear nonproliferation--[China] / Export controls--
[China]

TITLE:	Midnight never came.
SOURCE:	Bulletin of the atomic scientists, v. 51, Nov.-Dec. 1995: 16-27.
AUTHOR:	Moore, Mike.
LOCATION:	LRS95-9805 UA 17

NOTES: Discusses the "Bulletin Clock (first called 'The Clock of
Doom' and then 'The Doomsday Clock') entered folklore as a
symbol of nuclear peril and a constant warning that the
leaders of the United States and the Soviet Union had
better sit up and fly right." Considers the significance

in international relations of the clock, which was a symbol of the Bulletin of the Atomic Scientists.

SUBJECT(S): Nuclear weapons / Nuclear nonproliferation / Deterrence

---

TITLE: Table of contents page.
SOURCE: Bulletin of the atomic scientists, v. 51, Nov.-Dec. 1995.
LOCATION: LRS95-9802   Optical Disk

NOTES: Copy of the contents page of the journal annotated with document identification numbers (LRS numbers) to facilitate the retrieval of individual articles from the CRS optical disk system. To view the table of contents on the SCORPIO system, browse the title of the journal, select the set and display the most recent citations.

SUBJECT(S): Nuclear nonproliferation

TITLE: A cornerstone of world order: extending the NPT.
SOURCE: NATO review, v. 43, Sept. 1995: 21-26.
AUTHOR: Muller, Harald.
LOCATION: LRS95-9496   UA 17

NOTES: "Last May, delegates to the nuclear Non-Proliferation Treaty (NPT) review and extension conference adopted by consensus a proposal to extend the 25-year-old accord indefinitely; they also decided to strengthen the NPT review process, and they accepted a number of principles and objectives for nuclear non-proliferation and disarmament."

SUBJECT(S): Nuclear nonproliferation / Arms control agreements / Nuclear non-proliferation treaty (NPT)

TITLE: Against the spread of nuclear weapons: the safeguards system of the International Atomic Energy Agency.
SOURCE: NATO review, v. 43, Sept. 1995: 12-17.
AUTHOR: Blix, Hans.
LOCATION: LRS95-9493   QC 170 Gen.

NOTES: "The IAEA's role in verifying that a state's nuclear programmes are only being undertaken for peaceful purposes is a pivotal element of the non-proliferation regime. Verification measures-or safeguards as they are known--increase confidence among states, and can allay concerns which could provide the political motivation for the acquisition of nuclear weapons. While no diversion into nuclear weapons programmes of any significant quantity of material placed under safeguards has been detected, safeguards are now being strengthened in the light of recent experiences in such countries as Iraq and North

Korea. Looking to the future, the IAEA's experience and expertise could be used in verifying a comprehensive nuclear test ban, an agreed cut-off in the production of nuclear material for weapons use, or in verifying that nuclear material from dismantled weapons does not find its way into new nuclear weapons."

SUBJECT(S): Nuclear nonproliferation / Nuclear security measures / International Atomic Energy Agency

TITLE: Can the United States and Russia reshape the international strategic environment?
SOURCE: Comparative strategy, v. 14, 1995: 237-253.
AUTHOR: Benson, Sumner.
LOCATION: LRS95-9084 UV Gen.

NOTES: "The United States and Russia have agreed in principle that they should transform the strategic rivalry of the cold war into strategic cooperation in preventing proliferation of nuclear weapons, ballistic missiles, and advanced conventional weapons."

SUBJECT(S): Foreign relations--[U.S.]--Russia / Foreign relations--[Russia]--U.S. / Arms control / Strategic forces / Nuclear nonproliferation / International relations

TITLE: Nuclear threats today.
SOURCE: Medicine and war, v. 11, July-Sept. 1995: 29-103.
LOCATION: LRS95-8731 UC 650 A

NOTES: Contents.--The current status of the nuclear arms race, by Paul Rogers.--Fifty years after Nagasaki: Japan as plutonium superpower, by Shaun Burnie.--Current attitudes to the atomic bombings in Japan, by Kazuyo Yamane.--Illegal trafficking in nuclear fissile materials: likely customers and suppliers, by Jasjit Singh.--Do nuclear weapons have any rational utility? by Frank Barnaby.--The end of the beginning: progress towards the abolition of nuclear weapons, by Victor Sidel.--Nuclear weapons: the legality issue, by Rob Green.--A nuclear-weapon-free world: the essential lesson of Hiroshima, by Joseph Rotblat.

SUBJECT(S): Nuclear weapons / Arms control / Plutonium--[Japan] / Fissionable materials / Nuclear exports / Nuclear nonproliferation

TITLE:	Interim report of the Advisory Committee for Energy, Subcommittee on Nuclear Energy: multi-faceted measures under international cooperation aiming at securing safety of nuclear power generation in the Asian region. [Sogo Enerugi Chosakai Genshiryoku Bukai chukan hokokusho: kinrin Ajia chiiki ni okeru genshiryoku hatsuden no anzen kakuho o mezashita kokusai kyocho no motodeno tamenteki taisaku]
SOURCE:	Tokyo, The Ministry, 1995. ca. 52 p. in various pagings
AUTHOR:	Japan. Ministry of International Trade and Industry. Advisory Committee for Energy. Subcommittee on Nuclear Energy.
LOCATION:	LRS95-8674
NOTES:	Text in Japanese; citation title translated and abstract provided by the Japan Documentation Center. Asian countries including China, Korea, Taiwan and Indonesia have been recently active in constructing nuclear power stations. However, some nations are not prepared to maintain a domestic safety regulation system or accept an international treaty for ensuring safety of nuclear power generation. It is recognized that in case of atomic accidents occurring in a country, the damage can be stretched over the neighboring nations including Japan. The Subcommittee proposes Japan's possible cooperation for securing the safety over the Asian region through: 1) promoting information exchange among implementing countries of nuclear power generation on a private and governmental basis, and among countries exporting nuclear power equipment/material to Asian nations; 2) receiving more operators/engineers from those nations for education/training, dispatching experts to those nations for technical advice and providing safety management technology over the long term. The report points out that export restrictions should be strictly audited from the point of view of non-proliferation of nuclear weapons.
SUBJECT(S):	Nuclear energy policy--[Japan] / Nuclear nonproliferation--[Asia] / Nuclear power plants--[Japan]--Safety measures / Nuclear power plants--[Asia]--Safety measures / Nuclear power plant accidents--Prevention
ADDED ENTRY:	Japan. Tsusho Sangyosho. Sogo Enerugi Chosakai. Genshiryoku Bukai.

TITLE:	Managing nuclear proliferation in South Asia: an Indian view.	
SOURCE:	College Park, Center for International and Security Studies at Maryland, University of Maryland, 1995. 63 p. (CISSM papers 4)	
AUTHOR:	Chari, P. R. Hawes, John.	
LOCATION:	LRS95-8198	DS 350
NOTES:	"Initially it might be asserted that India's nuclear policy has evolved out of its conscious attempts, not unlike the	

United States, to balance its world interests--like universal peace, general and complete disarmament, an egalitarian international order--with the need to ensure its security, primarily from China and Pakistan." Includes an assessment by J. Hawes.

SUBJECT(S): Nuclear nonproliferation / Nuclear weapons--[India] / Collective security--[South Asia]

TITLE: Face proliferation directly; only penalties will deter Israel, India, Pakistan.
SOURCE: Defense news, Jan. 23, 1995: 2-3.
AUTHOR: Wolfsthal, Jon Brook.
LOCATION: LRS95-8094 UA 17

NOTES: "President Bill Clinton's administration has just repeated a common mistake of U.S. nonproliferation policy--ignoring proliferation in the hopes that it will eventually go away. Defense Secretary William Perry recently completed the first trip by a U.S. defense chief to Israel, India or Pakistan in several years. These three states are all capable of producing nuclear weapons and Israel, the most advanced of the three, may possess as many as 200 nuclear weapons."
This article was represented from the Nexis system.

SUBJECT(S): Nuclear nonproliferation

TITLE: Neo-nonproliferation.
SOURCE: Survival, v. 37, spring 1995: 66-85.
AUTHOR: Spector, Leonard S.
LOCATION: LRS95-8019 UA 17

NOTES: "In the strategic community, however, confidence in non-proliferation is waning and efforts are shifting away from preventing proliferation towards coping with its consequences. This new school of thought is becoming so pervasive and, . . . is so deliberately distancing itself from the traditional concept of non-proliferation that it deserves a label of its own--'neo-nonproliferation'. At a time when traditional non-proliferation efforts are having greater success than ever before and when broad historical trends presage further progress, neo-nonproliferation has begun to erode one of the most important pillars of American foreign policy."

SUBJECT(S): Nuclear nonproliferation / Collective security

TITLE:	Be very afraid: nukes, nerve gas and anthrax spores.
SOURCE:	New republic, v. 212, May 1, 1995: 19-20, 22-23, 26-27.
AUTHOR:	Wright, Robert.
LOCATION:	LRS95-7469 UA 17
NOTES:	Examines international arms control efforts and their effectiveness.
SUBJECT(S):	Arms control / Nuclear nonproliferation

TITLE:	Himalayan frontier under ominous cloud.
SOURCE:	Chicago tribune, Mar. 7, 1995: 1, 8.
AUTHOR:	Goozner, Merrill.
LOCATION:	LRS95-7457 DS 350
NOTES:	Examines ongoing tensions between India and Pakistan and considers the nature of this conflict given the undertone of nuclear weapons.
SUBJECT(S):	Foreign relations--[India]--Pakistan / Foreign relations--[Pakistan]--India / Nuclear nonproliferation

TITLE:	Putting nuclear nonproliferation control to the test.
SOURCE:	Aussenpolitik, v. 46, 1st quarter, 1995: 61-70.
AUTHOR:	Schilling, Walter.
LOCATION:	LRS95-7393 UA 17
NOTES:	"Treaty on the Nonproliferation of Nuclear Weapons (NP Treaty) is now due to be reviewed . . . . Examines the current NP configuration and assesses the prospects for the success of the forthcoming review conference."
SUBJECT(S):	Arms control agreements / Nuclear nonproliferation--Treaties / Treaty on the Nonproliferation of Nuclear Weapons (NPT)

TITLE:	End run around the NPT.
SOURCE:	Bulletin of the atomic scientists, v. 51, Sept.-Oct. 1995: 27-29.
AUTHOR:	Cabasso, Jacqueline.
LOCATION:	LRS95-7380 QC 170 U.S. D
NOTES:	Contends that the National Ignition Facility (NIF) will have impacts on proliferation.
SUBJECT(S):	Lasers--[U.S.] / Research and development facilities--[California] / Nuclear nonproliferation--[U.S.]

TITLE:	Table of contents page.
SOURCE:	Bulletin of the atomic scientists, v. 51, Sept.-Oct. 1995.
LOCATION:	LRS95-7377 Optical Disk
NOTES:	Copy of the contents page of the journal annotated with document identification numbers (LRS numbers) to facilitate the retrieval of individual articles from the CRS optical disk system. To view the table of contents on the SCORPIO system, browse the title of the journal, select the set and display the most recent citations.
SUBJECT(S):	Nuclear energy / Nuclear nonproliferation

TITLE:	Implications of the U.S.-North Korea nuclear agreement. Hearing, 103d Congress, 2nd session. Dec. 1, 1994.
SOURCE:	Washington, G.P.O., 1995. 97 p. (Hearing, Senate, 103d Congress, 2nd session, S. Hrg. 103-891)
AUTHOR:	U.S. Congress. Senate. Committee on Foreign Relations. Subcommittee on East Asian and Pacific Affairs.
LOCATION:	LRS95-7354 CONGRESSIONAL PUBLICATION RBC 7891
NOTES:	"The Agreed Framework with North Korea, was signed on October 21, 1994 . . . . This accord, to lift and replace North Korea's nuclear weapons facilities, involves a complex set of give and take steps over a lengthy period."
SUBJECT(S):	Nuclear weapons--[North Korea] / Arms control agreements--[U.S.]--North Korea / Arms control agreements--[North Korea]--U.S. / Nuclear nonproliferation / Nuclear research--[North Korea]

TITLE:	Nuclear policies in Northeast Asia. Edited by Andrew Mack.
SOURCE:	New York, United Nations, 1995. 265 p.
LOCATION:	LRS95-7335 LIMITED AVAILABILITY L SDI Loan
NOTES:	"UNIDIR/95/16" Partial contents.--Improving the global and regional nonproliferation regimes.--Northeast Asian security and nuclear proliferation.--Nuclear proliferation in Northeast Asia: regional perspectives.--Inter-relationships between regional and global approaches.
SUBJECT(S):	Nuclear nonproliferation--[Northeast Asia] / Nuclear weapons--[Northeast Asia] / Arms control / National security--[Northeast Asia] / United Nations Institute for Disarmament Research / North Korea
ADDED ENTRY:	Mack, Andrew.

TITLE:	Nuclear disarmament and non-proliferation in Northeast Asia.
SOURCE:	New York, United Nations, 1995. 83 p. (Research paper no. 33)
AUTHOR:	Han, Yong-Sup.
LOCATION:	LRS95-7334　　　　　　　　　　　　　　　　　　　DS 500
NOTES:	Partial contents.--Background.--China.--The Korean peninsula.--Japan.--Policy options.
SUBJECT(S):	Nuclear weapons--[Northeast Asia] / Nuclear nonproliferation--[Northeast Asia] / Arms control
ADDED ENTRY:	United Nations Institute for Disarmament Research.

TITLE:	[Japan's] fundamental plan for atomic energy development and use, 1995 [Heisei 7nendo genshiryoku kaihatsu riyo kihon keikaku]
SOURCE:	Tokyo, Prime Minister's Office, 1995. 40 p.
AUTHOR:	Japan. Prime Minister's Office.
LOCATION:	LRS95-7290　　　　　　　　　　　　　　　　　　　JDCTAF
NOTES:	Text in Japanese; citation title translated and abstract provided by the Japan Documentation Center. Atomic energy development and use is to be executed under the established basic concept: develop an atomic energy policy, as a nation that maintains its peaceful use; establish a well-coordinated power generation system with light water reactors; and steadily develop nuclear fuel recycling and versatile development/enhanced fundamental study of atomic energy technology. Particular areas are summarized: enhancement of the Non-Proliferation Treaty (NPT), measures for ensuring safety, information disclosure, promotion of back end measures, promotion of international cooperation, fostering and insurance of human resources, and others. The 1995 total budget for atomic energy development activities is 483.075 billion yen, up 5.7% from the previous year. The amounts are broken down by ministry or agency.
SUBJECT(S):	Nuclear energy policy--[Japan] / Nuclear nonproliferation
ADDED ENTRY:	Japan. Sorifu.

TITLE:	Rethinking the bomb, six part series.
SOURCE:	Washington post, Apr. 9, 1995: A1, A24-A25; Apr. 10: A1, A16-17; Apr. 11: A1, A16-A17; Apr. 12: A1, A28-A29; Apr. 13: A1, A26; Apr. 14: A1, A26-A27.
AUTHOR:	Coll, Steve. Ottaway, David B.
LOCATION:	LRS95-7112　　　　　　　　　　　　　　　　　　　UC 650 A
NOTES:	Contents.--A changing nuclear order.--U.S. focuses on threat of 'loose nukes'.--Secret visits helped define 3 powers' ties.--Trying to unplug the war machine.--New

threats create doubt in U.S. policy.--A hard sell for treaty renewal.
Washington post series

SUBJECT(S): Nuclear nonproliferation / Nuclear weapons / Arms control agreements

TITLE: Arms control. Prepared by the American Forces Information Service.
SOURCE: [Washington?] Current News Analysis & Research Service, 1995. 58 p. (U.S. Dept. of Defense. Current news, special edition, no. 95-6, July 1995)
LOCATION: LRS95-7088 UA 17

NOTES: A collection of editorials and newspaper articles

SUBJECT(S): Arms control / Nuclear weapons / Nuclear nonproliferation
ADDED ENTRY: U.S. Dept. of Defense.

TITLE: Nuclear safeguards and the International Atomic Energy Agency; summary.
SOURCE: Washington, Office of Technology Assessment, 1995. 22 p.
LOCATION: LRS95-6439 AVAILABLE FROM ISSUING AGENCY QC 170 Gen.

NOTES: Summary and press release material for the OTA study of Nuclear Safeguards and the International Atomic Energy Agency (LRS95-6438). "The International Atomic Energy (IAEA's) system of nuclear safeguards is intended to deter states from acquiring nuclear weapons by threatening to detect and expose such activities, such as the diversion of nuclear material from civil nuclear programs, if they occur. This Report analyzes what nuclear safeguards can and cannot be expected to accomplish and identifies areas where that system might be broadened and improved."

SUBJECT(S): Nuclear security measures / Nuclear nonproliferation / Fissionable materials--Safety measures / Nuclear weapons--Security measures / International control of nuclear power / International Atomic Energy Agency
ADDED ENTRY: U.S. Congress. Office of Technology Assessment.

TITLE: Nuclear safeguards and the International Atomic Energy Agency.
SOURCE: Washington, Office of Technology Assessment, for sale by the Supt. of Docs., G.P.O., 1995. 147 p.
LOCATION: LRS95-6438 LIMITED AVAILABILITY Optical Disk

NOTES: "This report analyzes what IAEA safeguards can and cannot be expected to accomplish, identifies areas where they might be broadened and improved, and presents options for doing so. It is the sixth publication of OTA's assessment on the proliferation of weapons of mass destruction."

SUBJECT(S): Nuclear security measures / Nuclear nonproliferation / Fissionable materials--Safety measures / Nuclear weapons--Security measures / International control of nuclear power / International Atomic Energy Agency
ADDED ENTRY: U.S. Congress. Office of Technology Assessment.

TITLE: My God...What have we done?
SOURCE: Newsweek, v. 126, July 24, 1995: 16-33, 36-41.
LOCATION: LRS95-6191   D 521 For.

NOTES: Describes the decision to drop the Atomic bomb on Hiroshima and Nagasaki and discusses the aftermath of the bombing. Includes an insert on the countries which have not signed the Nuclear Nonproliferation Treaty.

SUBJECT(S): Atomic bomb / World War II / Nuclear nonproliferation

TITLE: Nuclear nonproliferation: information on nuclear exports controlled by U.S.-EURATOM agreement: report to the Committee on Governmental Affairs, U.S. Senate. June 16, 1995.
SOURCE: Washington, G.A.O., 1995. 45 p.
AUTHOR: U.S. General Accounting Office.
LOCATION: LRS95-6104   AVAILABLE FROM ISSUING AGENCY   Optical Disk

NOTES: "GAO/RCED-95-168, B-261275"
"The U.S.-EURATOM agreement controls the exports of certain nuclear materials . . . . If a new agreement is not concluded before the expiration date, the export of these U.S. nuclear materials and components to EURATOM would be prohibited."

SUBJECT(S): Nuclear nonproliferation / Nuclear exports / EURATOM

TITLE: Koreapolitik.
SOURCE: Washington, National Defense University, Institute for National Strategic Studies, for sale by the Supt. of Docs., G.P.O., 1995. 4 p. (Strategic forum, no. 24, May 1995)
AUTHOR: Goodby, James. Drennan, William.
LOCATION: LRS95-6003   DS 920 C 3

NOTES: "The critical dimension of a peaceful solution to long-standing security issues in the Korean peninsula is a fruitful North-South dialogue. Hopes for such a dialogue were briefly raised by agreements reached by South and North Korea in 1991-92 and again by the U.S.-DPRK Agreed Framework of October 21, 1994 . . . . However, the inter-Korean security discussions have not progressed since then . . . and the confrontation between North and South Korea remains the most dangerous in the world." .650zx $Foreign relations[South Korea]North Korea

SUBJECT(S): Foreign relations--[North Korea]--South Korea / National security--[Korean Peninsula] / Nuclear nonproliferation / Nuclear weapons--[North Korea] / Conventional weapons--[Korean Peninsula] / Balance of power--[Korean Peninsula] / Korean Peninsula
ADDED ENTRY: National Defense University. Institute for National Strategic Studies.

TITLE: Proliferation and the nuclear disarmament process.
SOURCE: Energy policy, v. 23, March 1995: 195-207.
AUTHOR: Cochran, Thomas B.
LOCATION: LRS95-5849   UA 17

NOTES: "The greatest nuclear proliferation risk today arises from the lack of adequate physical protection, control and accounting of weapon usable materials in Russia. The most effective way to improve physical security and material accounting in Russia is through a cooperative effort to construct a comprehensive, non-discriminatory regime that ultimately would place all nuclear weapons and nuclear weapon usable materials under some form of multilateral monitoring. There is an urgent need for government action in states that now have significant programs involving the commercial use of nuclear weapon usable materials to defer further separation of plutonium until the global inventory of separated plutonium is significantly reduced and energy market conditions fully justify the added security risks of using plutonium in the civil fuel cycle."

SUBJECT(S): Nuclear nonproliferation / Arms control

TITLE: Indefinite extension--with increased accountability.
SOURCE: Bulletin of the atomic scientists, v. 51, July-Aug. 1995: 27-31.
AUTHOR: Epstein, William.
LOCATION: LRS95-5719   UA 17

NOTES: Discusses issues roused surrounding the extension of the NPT (Nuclear Nonproliferation Treaty).

SUBJECT(S): Nuclear nonproliferation / Arms control agreements

TITLE: Table of contents page.
SOURCE: Bulletin of the atomic scientists, v. 51, July-Aug. 1995.
LOCATION: LRS95-5717   Optical Disk

NOTES: Copy of the contents page of the journal annotated with document identification numbers (LRS numbers) to facilitate

the retrieval of individual articles from the CRS optical
disk system. To view the table of contents on the SCORPIO
system, browse the title of the journal, select the set and
display the most recent citations.

SUBJECT(S): Nuclear nonproliferation

TITLE: Proliferation and the North-South divide: the prospects for arms control.
SOURCE: International defense review--defense '95, 1995: 131-139.
AUTHOR: Stenhouse, Mark.
LOCATION: LRS95-5592 UV Gen.

NOTES: "Proliferation appears even more worrying when placed in the context of growing North-South divisions exacerbated by discrepancies of wealth and resources."

SUBJECT(S): Nuclear nonproliferation / Arms control agreements / Military policy

TITLE: Post-Cold War nuclear dangers: proliferation and terrorism.
SOURCE: Science, v. 267, Feb. 24, 1995: 1112-1114.
AUTHOR: Nuckolls, John H.
LOCATION: LRS95-5579 Q 170 Gen.

NOTES: Examines threats to stability caused by the proliferation of nuclear weapons and by terrorism. Views the role of science in decreasing these threats.

SUBJECT(S): Nuclear nonproliferation / Nuclear terrorism / Nuclear security measures

TITLE: U.S.-North Korea nuclear issues. Hearing, 104th Congress, 1st session. Jan. 19, 1995.
SOURCE: Washington, G.P.O., 1995. 88 p. (Hearing, Senate, 104th Congress, 1st session, S. Hrg. 104-5)
AUTHOR: U.S. Congress. Senate. Committee on Energy and Natural Resources.
LOCATION: LRS95-5516   CONGRESSIONAL PUBLICATION   RBC 8615

SUBJECT(S): Nuclear weapons--[North Korea] / Nuclear nonproliferation / Foreign relations--[U.S.]--North Korea / Foreign relations--[North Korea]--U.S.

TITLE: Fearsome security: the role of nuclear weapons.
SOURCE: Brookings review, v. 13, summer 1995: 24-27.
AUTHOR: May, Michael M.
LOCATION: LRS95-5344 UC 650 A

NOTES: "Winston Churchill warned at the beginning of the atomic age that safety could be the sturdy child of terror, and that we should not give up atomic weapons until we were sure and doubly sure that we had something better to take the place of terror in that respect."

SUBJECT(S): Nuclear weapons / Nuclear nonproliferation

TITLE: Bomb prevention vs. bomb promotion: exports in the 1990s. Hearing, 103d Congress, 2nd session. May 17, 1994.
SOURCE: Washington, G.P.O., 1995. 353 p. (Hearing, Senate, 103d Congress, 2nd session, S. Hrg. 103-1028)
AUTHOR: U.S. Congress. Senate. Committee on Affairs.
LOCATION: LRS95-5238   CONGRESSIONAL PUBLICATION   RBC 8459

SUBJECT(S): Export controls--[U.S.] / Nuclear nonproliferation / Nuclear exports--[U.S.] / Nuclear facilities--[Foreign]

TITLE: Weapons of mass destruction: new perspectives on counterproliferation. Edited by Stuart E. Johnson and William H. Lewis.
SOURCE: Fort Lesley J. McNair, Washington, D.C., National Defense University Press, for sale by the Supt. of Docs., G.P.O., 1995. 247 p.
LOCATION: LRS95-5141   LIMITED AVAILABILITY   L SDI Loan

NOTES: Partial contents.--Challenges of policy.--Regional challenges.--Preventive approaches.

SUBJECT(S): Weapons of mass destruction / Arms control / Nuclear nonproliferation / Military policy--[U.S.]
ADDED ENTRY: Johnson, Stuart E., Lewis, William H.

TITLE: Radical responses to radical regimes: evaluating preemptive counter-proliferation.
SOURCE: Washington, Institute for National Strategic Studies, National Defense University, for sale by the Supt. of Docs., G.P.O., 1995. 55 p. (McNair paper 41)
AUTHOR: Schneider, Barry R.
LOCATION: LRS95-5138   UA 17

NOTES: Partial contents.--Policy shift: the defense counter-proliferation initiative.--The spread of weapons of mass destruction.--History's lessons for preemptive counter-proliferation decisions.--Dealing with a potential "nuclear Hitler."--The proliferation challenge of North Korea.

SUBJECT(S): Nuclear nonproliferation / Arms control / Nuclear weapons / Weapons of mass destruction / Nuclear weapons--[North Korea]

TITLE:	Table of contents page.
SOURCE:	Bulletin of the atomic scientists, v. 51, May-June 1995.
LOCATION:	LRS95-4946　　　　　　　　　　　　　　　　Optical Disk
NOTES:	Copy of the contents page of the journal annotated with document identification numbers (LRS numbers) to facilitate the retrieval of individual articles from the CRS optical disk system. To view the table of contents on the SCORPIO system, browse the title of the journal, select the set and display the most recent citations.
SUBJECT(S):	Nuclear nonproliferation

TITLE:	Law and the H-bomb: strengthening the nonproliferation regime to impede advanced proliferation.
SOURCE:	Cornell international law journal, v. 28, winter 1995: 71-167.
AUTHOR:	Williamson, Richard L., Jr.
LOCATION:	LRS95-4829　　　　　　　　　　　　　　　　　　　UA 17
NOTES:	Partial contents.--Background to the advanced proliferation problem.--The nuclear nonproliferation regime and international law.--Legal instruments to restrain advanced proliferation. "This article examines the 'advanced proliferation' problem and the tools available to the international order to minimize it."
SUBJECT(S):	Nuclear nonproliferation / International law / Nuclear weapons / Arms control

TITLE:	North Korean military and nuclear proliferation threat: evaluation of the U.S.-DPRK Agreed Framework. Joint hearing before the Subcommittees on International Economic Policy and Trade and Asia and the Pacific of the Committee on International Relations, House of Representatives, 104th Congress, 1st session. Feb. 23, 1995.
SOURCE:	Washington, G.P.O., 1995. 136 p.
AUTHOR:	U.S. Congress. House. Committee on International Relations. Subcommittee on International Policy and Trade.
LOCATION:	LRS95-4694　　CONGRESSIONAL PUBLICATION　　　　RBC 8536
NOTES:	"The United States effort to address the regional security and nuclear proliferation threat from North Korea. Last October, the United States and North Korea signed an Agreed Framework."
SUBJECT(S):	Nuclear weapons--[North Korea] / Nuclear nonproliferation / Foreign relations--[U.S.]--North Korea / Foreign relations--[North Korea]--U.S. / National security--[Northeast Asia]
ADDED ENTRY:	U.S. Congress. House. Committee on International Relations. Subcommittee on Asia and the Pacific.

TITLE:	Dismantling the nuclear weapons legacy of the cold war.
SOURCE:	Washington, Institute for National Strategic Studies, 1995. 4 p. (Strategic forum no. 19)
AUTHOR:	Goodby, James E.
LOCATION:	LRS95-4287 UV For. CIS
NOTES:	"The process of dismantling former Soviet nuclear weapons systems, highly desirable to reduce a major threat to the United States, also generates new risks in the current security environment, including the risk of access to fissile materials by unauthorized individuals and criminal or terrorist organizations."
SUBJECT(S):	Nuclear nonproliferation--[CIS countries] / Nuclear weapons --[CIS countries] / Arms control--[CIS countries]

TITLE:	Working in the White House on nuclear nonproliferation and arms control: a personal report.
SOURCE:	F.A.S. public interest report, v. 48, Mar.-Apr. 1995: whole issue (8 p.)
AUTHOR:	von Hippel, Frank.
LOCATION:	LRS95-4282 UA 17
NOTES:	Former President advisor in the area of nuclear nonproliferation and arms control discusses his experiences."
SUBJECT(S):	Arms control / Nuclear nonproliferation

TITLE:	The politics of nuclear renunciation: the cases of Belarus, Kazakhstan, and Ukraine.
SOURCE:	Washington, Henry L. Stimson Center, 1995. 52 p. (Occasional paper no. 22, April 1995)
AUTHOR:	Potter, William C.
LOCATION:	LRS95-4159 UV For. CIS countries
NOTES:	Partial contents.--The Soviet nuclear weapons legacy.--From nuclear inheritance to renunciation.--The politics of nonproliferation.
SUBJECT(S):	Nuclear nonproliferation--[CIS countries] / Arms control / Nuclear weapons--[CIS countries]

TITLE:	The Future of the NPT.
SOURCE:	Arms control today, v. 25, Mar. 1995: whole issue (40 p.)
LOCATION:	LRS95-4155 UA 17
NOTES:	Contents.--The NPT: a global success story, by Spurgeon M. Keeny, Jr.--Extending the NPT: what are the options? by

George Bunn and John B. Rhinelander.--The nuclear-weapon states and article VI of the NPT, by Jack Mendelsohn and Dunbar Lockwood.--Key NPT conference issues, by Charles N. Van Dorn.--Summary and text of the NPT.

SUBJECT(S): Nuclear nonproliferation--Treaties / Arms control agreements / Nuclear weapons--Treaties / Non-Proliferation Treaty (NPT) / Treaty on the Nonproliferation of Nuclear Weapons

TITLE: Strategic implications of the U.S.-DPRK framework agreement.
SOURCE: Washington, U.S. Army War College, 1995. 33 p.
AUTHOR: Wilborn, Thomas L.
LOCATION: LRS95-4126   DS 930 E

NOTES: "After analyzing the agreement and its criticisms, the author argues that a major strategic consequence resulting from the process initiated by the Framework Agreement, in addition to halting the North Korean nuclear weapons program, is thrusting the United States into the middle of North-South confrontation on the Korean peninsula."

SUBJECT(S): Arms control agreements--[U.S.]--North Korea / Arms control agreements--[North Korea]--U.S. / Nuclear weapons--[North Korea] / Nuclear nonproliferation / Foreign relations--[Korean Peninsula]
ADDED ENTRY: Army War College (U.S.). Strategic Studies Institute.

TITLE: The origins, evolution and future of the North Korean nuclear program.
SOURCE: Korea and world affairs v. 19, spring 1995: 40-66.
AUTHOR: Mansourov, Alexandre Y.
LOCATION: LRS95-3953   DS 930 E

NOTES: "Traces the origins and evolution of the DPRK's nuclear program, and where it stands now .... Attempts to assess the DPRK's nuclear capabilities and intentions."

SUBJECT(S): Nuclear weapons--[North Korea] / Nuclear research--[North Korea] / Nuclear nonproliferation / Arms control

TITLE: A year of decision: arms control and non-proliferation in 1995.
SOURCE: U.S. Dept. of State Dispatch, v. 6, Feb. 6, 1995: 67-70.
AUTHOR: Lake, Anthony.
LOCATION: LRS95-3888   UA 17

NOTES: In this address, to the Carnegie Endowment for International Peace on Jan. 30, 1995, assistant to the President for National Security Affairs, discusses Clinton administration agenda for reducing the threat of nuclear weapons.

SUBJECT(S): Arm control agreements / Clinton Administration / Nuclear nonproliferation

TITLE:	Moonlighting by modem in Russia.
SOURCE:	U.S. News & world report, v. 118, Apr. 17, 1995: 45, 48.
AUTHOR:	Zimmerman, Tim. Pastenrak, Douglas.
LOCATION:	LRS95-3708 QC 170 For. CIS
NOTES:	Discusses how Russian scientists are using "GlasNet, a Moscow gateway to the worldwide Internet, "to communicate with parties interested in scientific and weapons information.
SUBJECT(S):	Brain drain--[Russia] / Nuclear weapons / Nuclear nonproliferation--[Russia] / Scientists in government--[Russia]

TITLE:	Critical mass.
SOURCE:	U.S. news & world report, v. 118, Apr. 17, 1995: 39-43, 45.
LOCATION:	LRS95-3707 UA 17
NOTES:	"The terrible dangers created by nuclear proliferation in the post-cold-war world have prompted growing calls from cold war hawks and doves alike for serious consideration of nuclear disarmament proposals. But technology is extremely difficult to undo, a truth that the creators of the bomb recognized from the start."
SUBJECT(S):	Nuclear nonproliferation

TITLE:	Middle Eastern security: prospects for an arms control regime.
SOURCE:	Contemporary security policy, v. 16, Apr. 1995: whole issue (194 p.)
AUTHOR:	Sandler, Shmuel.
LOCATION:	LRS95-3397 LIMITED AVAILABILITY L SDI Loan
NOTES:	Partial contents.--The global environment.--The international agenda.--The regional context.--Documents.
SUBJECT(S):	Nuclear nonproliferation / Arms control--[Middle East] / Ballistic missiles--[Middle East]
ADDED ENTRY:	Inbar, Efraim.

TITLE:	Clinton's "slow boat to Korea".
SOURCE:	Comparative strategy, v. 14, Jan.-Mar. 1995: 35-44.
AUTHOR:	Crouch, J. D.
LOCATION:	LRS95-3019 DS 930 E
NOTES:	"The Clinton administration's . . . policy has convinced the North Korean leadership that the longer it stalls, the more the concessions will be forthcoming . . . . If U.S. diplomacy in Northeast Asia is to work, it must be backed up by credible U.S. military power."
SUBJECT(S):	Foreign relations--[U.S.]--North Korea / Foreign relations--[North Korea]--U.S. / Nuclear weapons--[North Korea] / Nuclear nonproliferation / Arms control--[North Korea] / Clinton Administration

TITLE:	Science and technology issues.
SOURCE:	Forum for applied research and public policy, v. 10, spring 1995: 49-74.
LOCATION:	LRS95-2867 Q 170 For. CIS
NOTES:	Partial contents--Nuclear legacy of Soviet Union poses concerns, by William Potter.--Russia's nuke complex: a case for downsizing, by Oleg Bukharin.--Dismantling weapons: how will we know? by Mark A. Brown.--Legacy of cold war still plagues Russia, by Lydia Popova.--Chemical stockpiles seen as top priority, by Nikolai Plate, Yuly Kolbanovskii, Anatoly Ovsyannikov.
SUBJECT(S):	Nuclear weapons--[Russia] / Nuclear nonproliferation--[Russia] / Hazardous substances--[CIS countries]

TITLE:	Nuclear nonproliferation: time to make it permanent.
SOURCE:	Issues in science and technology, v. 11, spring 1995: 57-70.
AUTHOR:	Graham, Thomas, Jr.
LOCATION:	LRS95-2850 UA 17
NOTES:	Argues that "we must seize the opportunity to extend indefinitely the Nuclear Nonproliferation Treaty."
SUBJECT(S):	Arms control agreements / Nuclear nonproliferation

TITLE:	Nuclear proliferation factbook. Prepared for the Committee on Governmental Affairs, United States Senate by the Congressional Research Service, Library of Congress.
SOURCE:	Washington, G.P.O., 1995. 768 p. (Print, Senate, 103d Congress, 2d session, committee print S. Prt. 103-111)
AUTHOR:	U.S. Congress. Senate. Committee on Governmental Affairs.
LOCATION:	LRS95-1946 CONGRESSIONAL PUBLICATION RBC 8263

SUBJECT(S):	Nuclear nonproliferation--Treaties / International control of nuclear power / Nuclear security measures / Arms control agreements / Nuclear weapons / Uranium enrichment / Treaty on the Non-Proliferation of Nuclear Weapons / Atomic Energy Act / International Atomic Energy Agency
ADDED ENTRY:	Congressional Research Service.

TITLE:	Prospects for peace and stability in Northeast Asia: the Korean conflict.
SOURCE:	London, Research Institute for the Study of Conflict and Terrorism, 1995. 27 p. (Conflict studies 278, February 1995)
AUTHOR:	Gills, Barry K.
LOCATION:	LRS95-1905 F
NOTES:	"North Korea's refusal to allow access to its nuclear sites almost caused a world crisis."
SUBJECT(S):	Foreign relations--[U.S.]--North Korea / Foreign relations--[North Korea]--U.S. / Nuclear weapons--[North Korea] / Arms control--[Korean Peninsula] / Nuclear nonproliferation

TITLE:	The United States, Japan, and the future of nuclear weapons; report of the U.S.-Japan Study Group on Arms Control.
SOURCE:	Washington, Carnegie Endowment for International Pace, 1995. 181 p.
LOCATION:	LRS95-1846 LIMITED AVAILABILITY L SDI Loan
NOTES:	Contents.--Reducing nuclear weapons: is zero desirable.--The future of the non-proliferation treaty.--Obstacles to a comprehensive test ban.--Dismantling and disposing of nuclear weapons.--A weapons-grade fissile material cutoff.--Plutonium and nuclear power in Japan.--Arms control and tension reduction in East Asia.
SUBJECT(S):	Nuclear nonproliferation / Arms control--[U.S.] / Nuclear weapons tests / Military policy--[Japan]

TITLE:	Developments in North Korea. Hearing, 103d Congress, 2nd session. June 9, 1994.
SOURCE:	Washington, G.P.O., 1995. 32 p.
AUTHOR:	U.S. Congress. House. Committee on Foreign Affairs. Subcommittee on Asia and the Pacific.
LOCATION:	LRS95-1840 CONGRESSIONAL PUBLICATION RBC 8277
SUBJECT(S):	Nuclear weapons--[North Korea] / Nuclear nonproliferation / Arms control / Foreign relations--[U.S.]--North Korea / Foreign relations--[North Korea]--U.S.

TITLE:	Nuclear proliferation and nuclear entitlement.
SOURCE:	Ethics and international affairs, v. 9, 1995: 101-131.
AUTHOR:	Lee, Steven.
LOCATION:	LRS95-1834 UC 650 A
NOTES:	"Who has a right to the bomb? Is there an entitlement to nuclear weapons, and, if so, for whom? What can states justifiably do to stop other states from acquiring nuclear weapons? Is the Nonproliferation Treaty (NPT) discriminatory? These are moral questions, and, in addressing then in this essay, I will consider what ethics has to say about nuclear proliferation."
SUBJECT(S):	Nuclear nonproliferation / Arms control / Nuclear weapons / Ethics

TITLE:	Nuclear deterrence in a regional context.
SOURCE:	Santa Monica, Calif., Rand, 1995. 75 p.
AUTHOR:	Wilkening, Dean. Watman, Kenneth.
LOCATION:	LRS95-1295 UV U.S. A
NOTES:	Contents.--Why regional states might acquire nuclear weapons.--U.S. military capabilities.--An outline of U.S. strategies.--Implications for U.S. counterproliferation strategy.
SUBJECT(S):	Deterrence / Nuclear weapons / Nuclear nonproliferation / Military policy--[U.S.]
ADDED ENTRY:	Rand Corporation.

TITLE:	Beyond proliferation: the challenge of technology diffusion.
SOURCE:	Washington quarterly, v. 18, spring 1995: 183-202.
AUTHOR:	Moodie, Michael.
LOCATION:	LRS95-1273 Q 125 U.S. B
NOTES:	"The issue of the diffusion of militarily relevant technology in particular is introducing a new and contentious element into the international security arena. Disputes over export controls and the sharing of technology, for example, have emerged as major stumbling blocks in several arms-control forums."
SUBJECT(S):	Nuclear nonproliferation / Military research / Technology transfer / Technology and social problems

TITLE:	What vision for the nuclear future?
SOURCE:	Washington quarterly, v. 18, spring 1995: 127-142.
AUTHOR:	Daalder, Ivo H.
LOCATION:	LRS95-1264 UA 17

NOTES: Contends that "it is imperative that the United States and its allies succeed in their efforts to garner a substantial majority in favor of the NPT's indefinite and unconstitutional extension."

SUBJECT(S): Arms control / Nuclear nonproliferation

TITLE: Counterproliferation: putting new wine in old bottles.
SOURCE: Washington quarterly, v. 18, spring 1995: 143-154.
AUTHOR: Muller, Harald. Reiss, Mitchell.
LOCATION: LRS95-1263 UA 17

NOTES: "If the NPT is extended indefinitely and unconditionally, or even for an indefinite number of fixed periods, prospects for international peace and security will be significantly increased. A revised and expanded mission for counterproliferation would help to achieve this goal and win greater support for, as well as usefully reinforce, the international nonproliferation regime beyond 1995."

SUBJECT(S): Nuclear nonproliferation / Arms control

TITLE: Egypt needles Israel.
SOURCE: Bulletin of the atomic scientists, v. 51, Mar.-Apr. 1995: 11-13.
AUTHOR: Kumaraswamy, P. R.
LOCATION: LRS95-1100 UV For. Israel

NOTES: "For years Israel ignored Egypt's more private urgings that it accept some sort of international control over its nuclear facilities, and Egypt's increasingly voluble campaign reflects its displeasure with years of Israeli intransigence. As the April Nuclear Non-Proliferation Treaty (NPT) Extension Conference approaches, the rhetoric can be expected to heat up even more."

SUBJECT(S): Nuclear nonproliferation--[Israel]

TITLE: Non-proliferation treaty at 25.
SOURCE: CQ researcher, v. 5, Jan. 27, 1995: 73-96.
AUTHOR: Cooper, Mary H.
LOCATION: LRS95-977 UC 650 A

NOTES: "In April, negotiators from more than 160 countries will meet in New York to decide the fate of the 25-year-old Nuclear Non-Proliferation Treaty (NPT). Under the agreement's unusual terms, signatories must decide whether to extend it indefinitely, extend it for a specified period or abandon it. Supporters of a permanent NPT, including the United States, warn that instability in the post-Cold War era makes it all the more necessary to prevent rogue nations or terrorists from acquiring nuclear weapons. Critics, especially countries without nuclear weapons, say the treaty unfairly guarantees military superiority to the NPT's five-member 'nuclear club.'"

SUBJECT(S): Nuclear nonproliferation--Treaties / International control of nuclear power / Nuclear security measures--International cooperation / Nuclear weapons / Arms control / Nuclear nonproliferation--[U.S.]--Treaties / Nuclear nonproliferation--[Russia]--Treaties

TITLE: Getting down to business.
SOURCE: Bulletin of atomic scientists, v. 51, Jan.-Feb. 1995: 12-13.
AUTHOR: Lockwood, Dunbar.
LOCATION: LRS95-894 UV For. CIS

NOTES: "Three years ago, Congress passed the Nunn-Lugar bill to provide U.S. aid in denuclearizing and demilitaring the former Soviet Union. After a slow start, the program is now beginning."

SUBJECT(S): Nuclear nonproliferation

TITLE: Ukraine's decision to join the NPT.
SOURCE: Arms control today, v. 25, Jan.-Feb. 1995: 7-12.
AUTHOR: Garnett, Sherman W.
LOCATION: LRS95-806 DK 508

NOTES: "The key to success in U.S. policy toward Ukraine was the marriage of U.S. nuclear non-proliferation policy with a broad-based policy that supported economic and political reform and addressed Kiev's security concerns."

SUBJECT(S): Nuclear nonproliferation--[Ukraine] / Nuclear weapons--[Ukraine] / Arms control / Foreign relations--[U.S.]--Ukraine / Foreign relations--[Ukraine]--U.S.

TITLE: The United States and the prospects for NPT extension.
SOURCE: Arms control toady, Jan.-Feb. 1995: 3-6.
AUTHOR: Graham, Thomas, Jr.
LOCATION: LRS95-805 UA 17

NOTES: "The NPT's call for an end to the arms race has been met. Now, the race is on to bring down nuclear force levels as quickly, safely and securely as possible."

SUBJECT(S): Nuclear nonproliferation--Treaties / Arms control agreements / Arms control / Nuclear weapons tests

TITLE:	Rethinking nuclear proliferation.
SOURCE:	Washington quarterly, v. 18, winter 1995: 181-193.
AUTHOR:	Rathjens, George.
LOCATION:	LRS95-499 UC 650 A
NOTES:	"The Clinton administration is supporting the policies developed during those earlier years to deny potential proliferation access to materials and technology relevant to weapons acquisition--and delivery--and it will try, with considerable international support, to get the Nuclear Non-Proliferation Treaty (NPT) of 1970 extended well beyond 1995, when it comes up for review. It is also developing, within the Department of Defense, new 'counterproliferation' initiatives."
SUBJECT(S):	Nuclear nonproliferation / Clinton Administration

TITLE:	The technological promise of counterproliferation.
SOURCE:	Washington quarterly, v. 18, winter 1995: 153-166.
AUTHOR:	Pilat, Joseph F. Kirchner, Walter L.
LOCATION:	LRS95-497 UV Gen.
NOTES:	"With the end of the Cold War, there is a growing sense that the proliferation of weapons of mass destruction (WMD) and their means of delivery is dramatically worsening and that the existing nonproliferation regime is ill equipped to deal with the emerging threat. The U.S. Department of Defense (DOD) Bottom-Up Review of these weapons as the most urgent and direct threat to the security interests of the United States."
SUBJECT(S):	Nuclear nonproliferation / Military policy / Arms control

TITLE:	Nuclear rapprochement: Argentina, Brazil, and the nonproliferation regime.
SOURCE:	Washington Quarterly, v. 18, winter 1995: 107-122.
AUTHOR:	Redick, John R. Carasales, Julio C. Wrobel, Paulo S.
LOCATION:	LRS95-494 UC 650 A
NOTES:	"The approaching 1995 Conference on the Review and Extension of the Nuclear Non-Proliferation Treaty (NPT) provides an opportunity to assess the prospects for successful consolidation of the nuclear nonproliferation regime."
SUBJECT(S):	Nuclear nonproliferation / Arms control agreements

TITLE: Denial and deception practices of WMD proliferators: Iraq and beyond.
SOURCE: Washington quarterly, v. 18, winter 1995: 85-105.
AUTHOR: Kay, David A.
LOCATION: LRS95-493   DS 80

NOTES: With the end of the Cold War and the Soviet threat, a broad consensus seems to be forming that U.S. foreign and defense policies and budgets can now be premised on the absence of any major threat to the United States or its allies and on such timely detection of any newly developing threat that defensive measures can be deferred until after a threat emerges. The case of Iraq should offer a warning against too much optimism."

SUBJECT(S):   Persian Gulf War / Nuclear nonproliferation

TITLE: Jury-rigged, but working.
SOURCE: Bulletin of atomic scientists, v. 51, Jan.-Feb. 1995: 20-29.
AUTHOR: Albright, David. O'Neill, Kevin.
LOCATION: LRS95-13   UC 650 A

NOTES: Contends that "it wasn't just dumb luck that kept the number of nuclear weapon powers down to single digits, it was hard work and determination. In today's world--with excess weapon material and rogue states willing to use any means--the effort is as important as ever."

SUBJECT(S):   Nuclear nonproliferation / Nuclear weapons / Export controls

TITLE: South Korea's defense options regarding the North Korean nuclear crisis.
SOURCE: Korea & world affairs, v. 18, summer 1994: 322-346.
AUTHOR: Cho, Kap-Je.
LOCATION: LRS94-15731   DS 920 E

SUBJECT(S):   Nuclear weapons--[North Korea] / National defense--[South Korea] / Foreign relations--[U.S.]--North Korea / Foreign relations--[North Korea]--U.S. / Nuclear nonproliferation

TITLE: International Atomic Energy Agency (IAEA) report.
SOURCE: New York, United Nations, 1994. 12 p.
AUTHOR: United Nations. Security Council.
LOCATION: LRS94-15711   DS 930 E

NOTES: "S/1994/254, 4 March 1994"
              Discusses the IAEA findings concerning "the implementation of the Agreement between the Government of the Democratic People's Republic of Korea and the International Atomic

Energy Agency for the application of safeguards in
connection with the Treaty on the Non-Proliferation of
Nuclear Weapons."

SUBJECT(S): Nuclear security measures--[North Korea] / Nuclear
nonproliferation--[North Korea]

TITLE:	Toward a new Middle East: rethinking the nuclear question.
SOURCE:	Cambridge, Center for International Studies, Massachusetts Institute of Technology, 1994. 42 p.
AUTHOR:	Cohen, Avner.
LOCATION:	LRS94-15708 UV For. Middle East

NOTES: "Dramatic political developments have changed the face of
the Middle East during the last year: the Israeli-PLO Oslo
and Cairo accords that established the Palestinian National
Authority in the Gaza Strip and Jericho, and the Israeli-
Jordanian peace treaty signed on October 26, 1994 . . . .
What will be the impact of these fundamental changes on the
nuclear question in the Middle East . . . . Specifically,
are there circumstances under which Israel, the only de
facto nuclear state in the region, could be brought into
the nonproliferation regime? What about the Iran-Iraq
nuclear entanglement?"

SUBJECT(S): Nuclear weapons--[Middle East] / Peace negotiations--
[Middle East] / Nuclear nonproliferation--[Middle East] /
Arms control / Iraq / Iran / Israel

TITLE:	North Korea's nuclear development program and future.
SOURCE:	Korea & world affairs, v. 18, summer 1994: 273-300.
AUTHOR:	Kim, Hakjoon.
LOCATION:	LRS94-15575 DS 930 E

NOTES: "After Pyongyang announced its intention to withdraw from
the Nuclear Non-Proliferation Treaty (NPT) in March 1993,
the North Korea nuclear issue has grown into an
international controversy."

SUBJECT(S): Nuclear weapons--[North Korea]--Future / Nuclear facilities
--[North Korea] / Politics and government--[North Korea] /
Economic conditions--[North Korea] / Nuclear
nonproliferation / Arms control

TITLE:	Opportunities and challenges in Clinton's confidence-building strategy towards North Korea.
SOURCE:	Korean journal of defense analysis, v. 6, winter 1994: 145-156.
AUTHOR:	Niksch, Larry A.
LOCATION:	LRS94-15461 DS 930 C 2

NOTES: "The Clinton administration's decision in June 1994 to enter into comprehensive, high-level negotiations with North Korea's ended a period of two and one-half years of failed American diplomacy on the issue of the North Korean nuclear weapons program. The Bush and Clinton administrations had failed to gain a satisfactory outcome to the nuclear issue through a strategy that largely avoided sustained, face-to-face negotiations at a high level. In response to Jimmy Carter's visit to North Korea in June 1994, the Clinton administration decided to begin comprehensive talks with Pyongyang."

SUBJECT(S): Foreign relations--[U.S.]--North Korea / Foreign relations--[North Korea]--U.S. / Nuclear nonproliferation

TITLE: Confrontation and cooperation on the Korean peninsula: the politics of nuclear nonproliferation.
SOURCE: Korean journal of defense analysis, v. 6, winter 1994: 53-83.
AUTHOR: Koh, Byung Chul.
LOCATION: LRS94-15460 DS 930 E

NOTES: "North Korea's suspected nuclear weapons development program has raised the stakes for all of the principal players on or surrounding the Korean peninsula--the two Korean states, the United States, Japan, China, and Russia--albeit to varying degrees. It has also produced the paradoxical effect of simultaneously escalating confrontation and spawning cooperation among the erstwhile adversaries. All the players perceive the need to prevent confrontation from escalating to a conflict. They also have a common interest in elevating limited cooperation, as manifestation in dialogue and negotiation, to a higher plane where normal diplomatic relations and economic exchanges can materialize."

SUBJECT(S): Nuclear nonproliferation--[North Korea]

TITLE: Survival of rights under the Nuclear Non-Proliferation Treaty: withdrawal and the continuing right of international atomic energy agency safeguards.
SOURCE: Virginia journal of international law, v. 34, summer 1994: 749-830.
AUTHOR: Perez, Antonio F.
LOCATION: LRS94-15352 UA 17

NOTES: "This Article will then consider two different agruments in favor of a continuing right to inspections even against a state exercising its right to withdraw from the NPT to escape its safeguards obligations."

SUBJECT(S): Nuclear nonproliferation / Nuclear weapons / Nuclear weapons--[North Korea] / Inspection (Arms control)--[North Korea] / International Atomic Energy Agency / Treaty on Non-proliferation of Nuclear Weapons

TITLE: The United States and Pakistan: managing the relationship in a changing world.
SOURCE: Strategic studies, v. 17, autumn & winter 1994: whole issue (255 p.)
LOCATION: LRS94-15179   LIMITED AVAILABILITY   L SDI Loan

NOTES: Contents.--Future of US-Pakistan relations.--Resolving the Kashmir dispute.--Capping nuclear competition in South Asia.--Confidence building through crisis management and conventional arms balances.

SUBJECT(S): Foreign relations--[U.S.]--Pakistan / Foreign relations--[Pakistan]--U.S. / Nuclear weapons--[South Asia] / Nuclear nonproliferation / Balance of power--[South Asia] / Kashmir

TITLE: Confrontation and compromise over the North Korean nuclear problem.
SOURCE: Korean journal of defense analysis, v. 6, winter 1994: whole issue (374 p.)
LOCATION: LRS94-14978   L SDI Loan

NOTES: Partial contents.--Some consideration on resolving the North Korean nuclear question, by Ronald F. Lehman.--What can we do about nuclear forces in Northeast Asia? by Gerald Segal.--Confrontation and cooperation on the Korean Peninsula: the politics of nuclear nonproliferation, by Byung Chul Koh.--Korea as a pawn in the global nonproliferation conflict, by Kongdan Oh, and Ralph C. Hassig.--Should the United States supply light-water reactors to Pyongyang?

SUBJECT(S): Nuclear weapons--[North Korea] / Nuclear nonproliferation / Foreign relations--[U.S.]--North Korea / Foreign relations--[North Korea]--U.S. / Korean reunification

TITLE: Nuclear coexistence: rethinking U.S. policy to promote stability in an era of proliferation.
SOURCE: Montgomery, Ala., U.S. Air University, Air War College, for sale by the Supt. of Doc., G.P.O., 1994. 178 p. (Air War College studies in national security no. 1)
AUTHOR: Martel, William C. Pendley, William T.
LOCATION: LRS94-14895   LIMITED AVAILABILITY   L SDI LoanE885.M376
                                                              1994 (LC no.)

NOTES: "This study seeks to address the emerging incongruence between the proliferation of nuclear weapons and the U.S. policy for managing this process. American society and its

political leadership must accept the need to adapt its
policy to the rapidly-changing circumstances in nuclear
proliferation. For at least two decades, the process of
nuclear proliferation continued unabated, with the
emergence of new nuclear powers, including India, Israel,
and Pakistan. Since 1992, deep concerns about the
emergence of North Korea as a nuclear power have provoked a
protracted diplomatic crisis between the South Korean-
United States alliance and North Korea. Further, the
dissolution of the Soviet Union created three additional
'instant' nuclear powers--Ukraine, Kazakhstan, and
Belarus."

SUBJECT(S): Nuclear nonproliferation
ADDED ENTRY: Air University (U.S.). Air War College.

TITLE: South Asian nuclear proliferation: a case for co-operation under anarchy?
SOURCE: Contemporary security policy, v. 15, Dec. 1994: 199-227.
AUTHOR: Gallagher, Nancy W. Bajpai, Kanti.
LOCATION: LRS94-14610   DS 350

NOTES: Applies traditional nonproliferation theories to events in South Asia.

SUBJECT(S): Nuclear nonproliferation--[South Asia] / Collective security--[South Asia]

TITLE: US non-proliferation policy: global regimes and regional realities.
SOURCE: Contemporary security policy, v. 15, Dec. 1994: 128-146.
AUTHOR: Steinberg, Gerald M.
LOCATION: LRS94-14503   UA 17

NOTES: "The purpose of this paper is to examine the role of global institutions and the universal rules of behaviour in the existing non-proliferation regime and American policy, and to analyse the significance of this policy in several regional contexts, with a particular emphasis on the Middle East."

SUBJECT(S): Nuclear nonproliferation

TITLE: Security issues in the handling and disposition of fissionable material.
SOURCE: Contemporary security policy, v. 15, Dec. 1994: 1-29.
AUTHOR: Abrams, Herbert L. Pollak, Daniel.
LOCATION: LRS94-14498   Q 170 U.S. A

NOTES: Examines "the nature of the risks and the various measures that have evolved in the United States to protect and

Nuclear Proliferation: An Annotated Biography 191

                account for nuclear materials .... Using the US system as a basis for comparisons, reviews the situation in Russia, and indicates ways in which their security measures fall below US standards .... Examines the process of warhead dismantlement and fissile material disposal, with specific attention to the steps that will pose challenges to security and control .... [Identifies] additional measures required to improve security in the US and Russia so as to minimize proliferation risks during warhead dismantlement and the disposal of material."

SUBJECT(S): Hazardous substances--[U.S.] / Fissionable materials--[U.S.] / Nuclear weapons / Nuclear nonproliferation

---

TITLE: Nuclear proliferation: the post-cold-war challenge.
SOURCE: Ithaca, N.Y., Foreign Policy Association, 1994. 72 p. (Headline series no. 303)
AUTHOR: Bee, Ronald J.
LOCATION: LRS94-14470   UA 17

NOTES: Contents.--The dawn of nuclear proliferation.--Three nuclear races.--The nonproliferation regime.--Nuclear hot spots: former Soviet Union; North Korea; Iraq; Iran and Israel; India and Pakistan.

SUBJECT(S): Nuclear weapons / Nuclear nonproliferation / Arms control

---

TITLE: Toward a new Middle East: rethinking the nuclear question.
SOURCE: Cambridge, Center for International Studies, Massachusetts Institute of Technology, 1994. 42 p. (DACS working paper (Nov. 1994))
AUTHOR: Cohen, Avner.
LOCATION: LRS94-14291   F Library

NOTES: "WP 94-3"
Examines the Middle East nuclear situation in light of "the experience of Iraq, North Korea, South Africa, and Latin America."

SUBJECT(S): Nuclear nonproliferation--[Middle East] / Arms control / Nuclear energy

---

TITLE: North Korea's nuclear development program and future.
SOURCE: Korea & world affairs, v. 18, summer 194: 273-300.
AUTHOR: Kim, Hakjoon.
LOCATION: LRS94-14107   DS 930 E

NOTES: "This article intends to assess the issue of North Korea's nuclear development program from the perspective of the argument that the North Korean regime is bound to collapse ... addresses the process of the emergence of the North

Korean nuclear development issue, then assesses the series
of dialogues concerning the issue between North Korea and
the United States and North Korea and the IAEA."

SUBJECT(S): Nuclear weapons--[North Korea] / Politics and government--
[North Korea] / Nuclear nonproliferation / Arms control--
[North Korea] / Foreign relations--[U.S.]--North Korea /
Foreign relations--[North Korea]--U.S. / International
Atomic Energy Agency (IAEA)

TITLE: China's leverages over North Korea.
SOURCE: Korea & world affairs, v. 18, summer 1994: 233-249.
AUTHOR: Han, Yong-Sup.
LOCATION: LRS94-14105   JX 1428 For. China E 3

NOTES: "The purpose of this article is to identify to what extent
             China can exert leverage over North Korea, and what factors
             China will take into consideration when requested to
             participate in sanctions against North Korea."

SUBJECT(S): Foreign relations--[China]--North Korea / Foreign relations
--[North Korea]--China / Nuclear weapons--[North Korea] /
Nuclear nonproliferation

TITLE: North Korea--to the brink and back.
SOURCE: Jane's intelligence review yearbook, 1994-95: 131-135.
AUTHOR: Gerardi, Greg. Plotts, James A.
LOCATION: LRS94-14025   DS 930 E

NOTES: "1994 was a tumultuous year of succeeding international and
             national crises for North Korea. The nuclear crisis formed
             the centerpiece."

SUBJECT(S): Nuclear weapons--[North Korea] / Nuclear nonproliferation /
Nuclear facilities--[North Korea] / Politics and government
--[North Korea] / Inspection (Arms control)--[North
America] / International Atomic Energy Agency

TITLE: Nuclear non-proliferation via a comprehensive test ban?
SOURCE: Medicine and war, v. 10, Oct.-Dec. 1994: 314-318.
AUTHOR: Doucet, Ian.
LOCATION: LRS94-13994   UA 17

NOTES: "The Council for Arms Control held a seminar on the
             negotiations toward a test ban . . . . Efforts towards a
             Comprehensive Test Ban Treaty (CTBT) and the issue of
             preventing nuclear proliferation have received much greater
             public and political attention in the United States than in
             Britain . . . . The seminar left the impression, that a
             CTBT would be insufficient assurance of the nuclear weapons
             states' future compliance with Article VI of the NPT."

SUBJECT(S): Nuclear nonproliferation / Nuclear weapons tests / Arms control / Comprehensive Test Ban Treaty (Proposed) / Council for Arms Control

TITLE: The proliferation of nuclear, biological and chemical weapons.
SOURCE: Jane's intelligence review yearbook, 1994-1995: 16-19.,
AUTHOR: Latter, Richard.
LOCATION: LRS94-13943   UA 17
NOTES: "An intensification of efforts to prevent the proliferation of weapons of mass destruction (WMD). Much has been achieved .... Unfortunately, this progress has been paralleled by more negative developments .... Is an open question whether the positive or the negative trends will predominate."
SUBJECT(S): Weapons of mass destruction / Arms control / Nuclear nonproliferation / Nuclear weapons / Biological weapons / Chemical weapons

TITLE: Plutonium disposition. Hearing, 103d Congress, 2nd session. May 26, 1994.
SOURCE: Washington, G.P.O., 1994. 104 p. (Hearing, Senate, 103d Congress, 2nd session, S. Hrg. 103-857)
AUTHOR: U.S. Congress. Senate. Committee on Energy and Natural Resources.
LOCATION: LRS94-13559   CONGRESSIONAL PUBLICATION   RBC 7949
SUBJECT(S): Plutonium--[U.S.] / Plutonium--[CIS countries] / Nuclear nonproliferation / Nuclear security measures--[U.S.] / Nuclear weapons--[U.S.] / Nuclear weapons--[CIS countries] / Nuclear reactors--[U.S.]--Research

TITLE: Nuclear nonproliferation: U.S. international nuclear materials tracking capabilities are limited; report to congressional requesters. Dec. 27, 1994.
SOURCE: Washington, G.A.O., 1994. 27 p.
AUTHOR: U.S. General Accounting Office.
LOCATION: LRS94-13178   AVAILABLE FROM ISSUING AGENCY   Optical Disk
NOTES: "GAO/RCED/AIMD-95-5, B-259533"
"How the United States tracks its exported civilian (nondefense-use) nuclear materials and ensures their physical protection."
SUBJECT(S): Fissionable materials--[U.S.]--Security measures / Nuclear nonproliferation / Nuclear exports--[U.S.] / Export controls

TITLE:	The NPT renewal conference.
SOURCE:	International security, v. 19, summer 1994: 41-71.
AUTHOR:	Simpson, John. Howlett, Darryl.
LOCATION:	LRS94-13142 UA 17
NOTES:	"The greatest uncertainty in the nuclear non-proliferation area remains the outcome of the 1995 Conference to review and extend the Treaty on the Non-Proliferation of Nuclear Weapons (NPT). This Conference, to be convened in New York from April 17 to May 12, 1995 . . . . Since it entered into force on March 5, 1970, the NPT has become the mainstay of the nuclear non-proliferation regime."
SUBJECT(S):	Nuclear nonproliferation--Treaties / Arms control agreements / Nuclear weapons

TITLE:	The development of nonproliferation export control in Russia.
SOURCE:	World affairs, v. 157, summer 1994: 3-18.
AUTHOR:	Beck, Michael. Bertsch, Gary. Khripunov, Igor.
LOCATION:	LRS94-12986 HF 1455
NOTES:	"The problems of post-communism [include] the threat of weapons proliferation from the former Soviet Union, and especially from Russia . . . . Russia is attempting to develop and implement non-proliferation export controls in line with Western standards."
SUBJECT(S):	Nuclear nonproliferation--[Russia] / Export controls--[Russia]

TITLE:	Let 'em have nukes.
SOURCE:	New York times magazine, Nov. 13, 1994: 56-57.
AUTHOR:	Bandow, Doug.
LOCATION:	LRS94-12786 UC 650 A
NOTES:	"Washington's pursuit of a world without nuclear weapons is a dangerous fantasy."
SUBJECT(S):	Nuclear weapons / Nuclear nonproliferation / Arms control / Military policy--[U.S.] / Foreign relations--[U.S.] / North Korea

TITLE:	North Koran time bomb: can sanctions defuse it? A review of international economic sanctions as an option.
SOURCE:	Georgia journal of international and comparative law, v. 24, fall 1994: 307-346.
AUTHOR:	Pires, Jeong Hwa.
LOCATION:	LRS94-12675 DS 930 E

NOTES: "For more than two years, North Korea has played games with the West over its nuclear program by promising access to International Atomic Energy Agency (IAEA) inspectors as is required under the Nuclear Non-Proliferation Treaty (NPT), ... The danger presented by the North Koran nuclear program . . . extends to the rest of the world because of North Korea's history of exporting weapons technology . . . . This Note will review the legal background for the imposition of universal economic sanctions against a state and discuss the special circumstances which will affect the outcome of the economic sanctions currently being considered."

SUBJECT(S): Nuclear weapons--[North Korea] / Sanctions (International law)--[North Korea] / Nuclear nonproliferation--[North Korea] / Nuclear facilities--[North Korea]

TITLE: Comprehensive negotiations with North Korea: a viable alternative for a failed U.S. strategy.
SOURCE: Korea & world affairs, v. 18, summer 1994: 250-272.
AUTHOR: Niksch, Larry A.
LOCATION: LRS94-12085   DS 930 C 2

SUBJECT(S): Foreign relations--[U.S.]--North Korea / Foreign relations--[North Korea]--U.S. / Clinton Administration / Nuclear weapons--[North Korea] / Nuclear nonproliferation / Foreign relations--[Northeast Asia]

TITLE: North Korea's nuclear weapons programme: genesis, motives, implications.
SOURCE: Aussenpolitik, v. 45, no. 4, 1994: 354-363.
AUTHOR: Maull, Hanns W.
LOCATION: LRS94-12075   DS 930 E

NOTES: "The appeal of these weapons, which have become by and large uninteresting for the established nuclear-weapon powers and are thus experiencing a process of down-scaling, has increased for the previous 'have-nots.' One of the latter is Communist North Korea, a country which is moving into a critical internal and external phase."

SUBJECT(S): Nuclear weapons--[North Korea] / Nuclear nonproliferation / Arms control

TITLE: Containment of nuclear risks.
SOURCE: Aussenpolitik, v. 45, no. 4, 1994: 346-353.
AUTHOR: Dregger, Alfred.
LOCATION: LRS94-12074   UC 650 A

NOTES: "One of the major emergent challenges for the international community is the further proliferation of nuclear weapons.

                     The conflicts with North Korea have highlighted the issue."
SUBJECT(S):   Nuclear weapons / Nuclear nonproliferation / Arms control

TITLE: Nuclear safeguards and security in the former Soviet Union.
SOURCE: Survival, v. 36, winter 1994-95: 53-72.
AUTHOR: Bukharin, Oleg.
LOCATION: LRS94-11844   DK 350 E

NOTES: "The importance of Russia and other former republics for
              the global fissile-material-control. The republics, and
              especially Russia, can play a crucial role in enforcing the
              international fissile-material-control regime by
              participating in global non-proliferation efforts."

SUBJECT(S):   Fissionable materials--[Russia]--Safety measures / Nuclear
              weapons--[Russia]--Security measures / Nuclear
              nonproliferation / Port controls--[Russia]

TITLE: The IAEA's dirty little secret.
SOURCE: International economy, v. 8, Sept.-Oct. 1994: 52-55.
AUTHOR: Gaffney, Frank J., Jr.
LOCATION: LRS94-11547   QC 170 Gen.

NOTES: Contends that "the International Atomic Energy Agency is
              increasingly incapable of preventing nuclear
              proliferation."

SUBJECT(S):   Nuclear nonproliferation / Inspection (Arms control) /
              Nuclear weapons / Nuclear energy / International Atomic
              Energy Agency--Evaluation

TITLE: Phase out the bomb.
SOURCE: Foreign policy, no. 97, winter 1994-95: 79-96.
AUTHOR: Blechman, Barry M.   Fisher, Cathleen S.
LOCATION: LRS94-11519   UA 17

NOTES: Concludes that "if Americans are serious about stopping the
              spread of weapons of mass destruction, then they must make
              every effort to strengthen the nonproliferation norm and to
              delegitimate the weapons. A clear commitment to their
              elimination would be an important first step toward that
              objective. It would end the glaring hypocrisy that
              cripples the current U.S. nonproliferation approach."

SUBJECT(S):   Atomic bomb / Arms control / Nuclear nonproliferation

TITLE:	The impact of nuclear proliferation: the case of Algeria, 2003.
SOURCE:	Alexandria, Va., Center for Naval Analyses, 1994. 69 p.
AUTHOR:	Gonzalez, Iris M. Walne, George. Warren, James.
LOCATION:	LRS94-11426 DT 295
NOTES:	"CRM94-40/September 1994"
"This research memorandum is part of a CNA-sponsored study that identifies the policy and force implications for the United States of nuclear weapon proliferation in distant Third World areas. This particular memorandum identifies implications of nuclear possession by Algeria in 2003 for policies and programs affecting the design, organization, location, and employment of future U.S. forces."
SUBJECT(S):	Nuclear nonproliferation--[Algeria]--Scenarios / Collective security

TITLE:	The impact of nuclear proliferation: the case of Iran, 1997.
SOURCE:	Alexandria, Va., Center for Naval Analyses, 1994. 77 p.
AUTHOR:	Gonzalez, Iris M. Walne, George. Warren, James.
LOCATION:	LRS94-11425 F Library
NOTES:	"CRM93-231/September 1994"
"This research memorandum is part of a CNA-sponsored study that identifies the policy and force implications for the United States of nuclear weapons proliferation in distant Third World areas. In this particular memorandum, we examine a scenario that illustrates a political and military standoff between Iran and the United States assumed to take place in 1997."
SUBJECT(S):	Nuclear nonproliferation--[Iran]--Scenarios / Foreign relations--[U.S.]--Iran / Foreign relations--[Iran]--U.S.

TITLE:	Germinating technology feeds a-weapon scenario.
SOURCE:	National defense, v. 79, Oct. 1994: 18-19.
AUTHOR:	Evancoe, Paul R.
LOCATION:	LRS94-11233 UC 650 A
NOTES:	"The mounting capabilities by rogue nations to build nuclear weapons are rendering international non-proliferation treaties increasingly ineffective in containing the spread of nuclear weapon technology, say experts. Proliferation of nuclear weapons technology and the special nuclear material required to build a nuclear weapon has remained a top national security concern for the United States and its allies."
SUBJECT(S):	Nuclear nonproliferation / Plutonium / Nuclear weapons

TITLE: Nuclear proliferation and counter-proliferation: policy issues and debates.
SOURCE: Mershon international studies review, v. 38, Oct. 1994: 209-234.
AUTHOR: Schneider, Barry R.
LOCATION: LRS94-10817   UC 650 A

NOTES: "This review provides a map of recent approaches to and issues within the literature on nuclear proliferation and counterproliferation. It identifies and explains six approaches to proliferation theory and policy."

SUBJECT(S): Nuclear weapons / Nuclear nonproliferation

TITLE: Non-proliferation, self-defense, and the Korean crisis.
SOURCE: Vanderbilt journal of transnational law, v. 27, Oct. 1994: 603-634.
AUTHOR: Newcomb, Mark E.
LOCATION: LRS94-10589   DS 930 E

NOTES: "This Article investigates the history of the Korean crisis and places North Korea's attempt to withdraw from the Treaty on the Non-Proliferation of Nuclear Weapons in the context of the international non-proliferation regime and policy.... The traditional concept of self-defense is inadequate to deal with the problem presented by a nuclear threat."

SUBJECT(S): Nuclear weapons--[North Korea] / Nuclear nonproliferation / Arms control / Collective security / Foreign relations--[U.S.]--North Korea / Foreign relations--[North Korea]--U.S.

TITLE: The unlawful plutonium alliance: Japan's supergrade plutonium and the role of the United States.
SOURCE: Amsterdam, Greenpeace International, 1994. ca. 100 p. in various pagings
AUTHOR: Greenpeace International.
LOCATION: LRS94-10399   QC 170 Gen.

NOTES: "The focus of this paper is the little-publicized plutonium alliance between the United States and Japan. Documents obtained through the U.S. Freedom of Information Act reveal that rather than discouraging weapons capable nuclear programmes in the region, the United States has been actively assisting Japan acquire advanced plutonium technologies. Developed at a cost of billions of dollars to the U.S. taxpayer, this civil-military technology is to be incorporated into a new Japanese plutonium reprocessing facility. Once operational the plant will give Japan access to the highest quality plutonium ideal for sophisticated nuclear weapons."

SUBJECT(S): Plutonium--[Japan] / Reactor fuel reprocessing--[Japan] / Nuclear nonproliferation--[U.S.] / Nuclear nonproliferation--[Japan] / Technology transfer--[U.S.]--Japan

TITLE: The silent threat--nuclear reactor safety in Eastern Europe and the CIS.
SOURCE: Jane's intelligence review, v. 6, Aug. 1994: 358-362.
AUTHOR: Halverson, Thomas.
LOCATION: LRS94-10393 QC 170 For. CIS countries

NOTES: Examines the safety and security of Soviet-designed nuclear power reactors still in operation. Also contemplates the proliferation risks associated with uranium stockpiles in the Ukraine.

SUBJECT(S): Nuclear reactors--[CIS countries]--Safety measures / Nuclear reactors--[Eastern Europe]--Safety measures / Nuclear security measures--[Eastern Europe] / Nuclear security measures--[CIS countries] / Uranium--[CIS countries] / Nuclear nonproliferation--[U.S.] / Nuclear nonproliferation--[CIS countries] / Reactor fuel reprocessing--[CIS countries]

TITLE: Proliferation and the former Soviet Union.
SOURCE: Washington, Office of Technology Assessment, for sale by the Supt. of Docs., G.P.O., 1994. 92 p.
LOCATION: LRS94-10374 AVAILABLE FROM ISSUING AGENCY QC 170 For. CIS countries

NOTES: "OTA-ISC-605, September 1994"
""Although greeted with relief by many both within and without the Soviet Union, that nation's collapse in 1991 gave rise to serious concerns over the loss of responsible state control over Soviet nuclear weapons, the materials used to build them, and the expertise, information, and technology required to produce nuclear or other weapons of mass destruction . . . . There are measures that the United States can take and is taking to reduce the chances that the Soviet breakup will aggravate the proliferation of weapons of mass destruction. These measures are described and analyzed."

SUBJECT(S): Nuclear nonproliferation--[CIS countries] / Nuclear weapons--[CIS countries] / International control of nuclear power--[U.S.]
ADDED ENTRY: U.S. Congress. Office of Technology Assessment.

TITLE:	North Korea--where to from here?
SOURCE:	Jane's intelligence review, v. 6, Aug. 1994: 374-383.
AUTHOR:	Segal, Gerald. Mussington, David.
LOCATION:	LRS94-10190 DS 930 E
NOTES:	"Attempts to undo North Korea's earlier flagrant violations of the Non-Proliferation Treaty (NPT) and rulings from the IAEA."
SUBJECT(S):	Nuclear weapons--[North Korea] / Inspection (Arms control)--[North Korea] / Nuclear nonproliferation / Arms control / Nuclear facilities--[North Korea]

TITLE:	NATO and the coming proliferation threat: view from NATO.
SOURCE:	Comparative strategy, v. 13, July-Sept. 1994: 313-320.
AUTHOR:	Ruhle, Michael.
LOCATION:	LRS94-10153 JX 4005 C
NOTES:	"While NATO's active role in nonproliferation is very recent and not yet fully defined, this paper has attempted to indicate those areas where a NATO role could be developed in the longer term. NATO's nonproliferation activities are intended to support existing efforts."
SUBJECT(S):	Nuclear nonproliferation / Military policy--[NATO countries]

TITLE:	British, French, and Chinese nuclear forces: implications for arms control and nonproliferation.
SOURCE:	College Park, Md., Center for International and Security Studies at Maryland, 1994. 16 p.
AUTHOR:	Norris, Robert S.
LOCATION:	LRS94-10133 UC 650 A
NOTES:	"Although the geopolitical situation has changed drastically in recent years, none of the five acknowledged nuclear powers has decided to dismantle its nuclear arsenal . . . All five have decided that continued possession of nuclear weapons will remain a part of their security policies. In fact, in all five countries, modernization programs proceed."
SUBJECT(S):	Nuclear weapons--[Great Britain] / Nuclear weapons--[France] / Nuclear weapons--[China] / Arms control / Nuclear nonproliferation
ADDED ENTRY:	University of Maryland, College Park. Center for International Security Studies at Maryland.

TITLE:	Stand and be counted.
SOURCE:	Bulletin of the atomic scientists, v. 50, Nov.-Dec. 1994: 15-23.
AUTHOR:	Epstein, William.
LOCATION:	LRS94-9886 UC 650 A
NOTES:	"After 25 years, the Nuclear Non-Proliferation Treaty is up for debate . . . . [Presents] two views of the conditions for its extension, and a look at a closely related issue."
SUBJECT(S):	Nuclear nonproliferation / Arms control agreements / Nuclear Nonproliferation Treaty

TITLE:	Sweden without the bomb: the conduct of a nuclear-capable nation without nuclear weapons.
SOURCE:	Santa Monica, Calif., Rand Corporation, 1994. 273 p. (MR-460)
AUTHOR:	Cole, Paul M.
LOCATION:	LRS94-9852 LIMITED AVAILABILITY L SDI Loan
NOTES:	Partial contents.--Sweden's nuclear weapon research.--Sweden's delivery systems and efforts to obtain dual-capable U.S. missiles.--Sweden's doctrine and strategy.--Why Sweden did not acquire nuclear weapons.--Relevance to contemporary cases.
SUBJECT(S):	Nuclear research--[Sweden] / Nuclear weapons--[Sweden]--Research / National security--[Sweden] / Nuclear nonproliferation
ADDED ENTRY:	Rand Corporation.

TITLE:	Proliferation and the former Soviet Union.
SOURCE:	Washington, Office of Technology Assessment, for sale by the Supt. of Docs., G.P.O., 1994. 92 p.
LOCATION:	LRS94-9729 AVAILABLE FROM ISSUING AGENCY UC 650 A
NOTES:	Partial contents.--Part I: executive summary.--Part II: proliferation threats and responses; threats to international nonproliferation regimes; blocking access to nuclear weapons, materials and expertise.--Part III: the individual nuclear inheritor states; Belarus; Kazakhstan; Russia; Ukraine.
SUBJECT(S):	Nuclear weapons--[Russia] / Nuclear weapons--[Belarus] / Nuclear weapons--[Ukraine] / Nuclear weapons--[Kazakhstan] / Nuclear nonproliferation
ADDED ENTRY:	U.S. Office of Technology Assessment.

TITLE:	Stemming the plutonium tide: limiting the accumulation of excess weapon-usable nuclear materials. Hearing, 103d Congress, 2nd session. Mar. 23, 1994.
SOURCE:	Washington, G.P.O., 1994. 138 p.
AUTHOR:	U.S. Congress. House. Committee on Foreign Affairs. Subcommittee on International Security, International Organizations and Human Rights.
LOCATION:	LRS94-9728 CONGRESSIONAL PUBLICATION RBC 7228
NOTES:	"What can be done to reduce the further accumulation of plutonium in the years ahead? And, secondly, what are the prospects for ensuring that surplus plutonium does not fall into the wrong hands?"
SUBJECT(S):	Plutonium / Arms control / Fissionable materials / Nuclear nonproliferation

TITLE:	North Korea's future: dynamism of economic reform and the nuclear option.
SOURCE:	[New York] East Asian Institute, Columbia University, 1994. 36 p.
AUTHOR:	Tanaka, Yoshikazu.
LOCATION:	LRS94-9600 DS 930 E
NOTES:	"Can North Korea survive this crisis? To answer this question one must look at North Korea's intention when it announced its withdrawal from the NPT. In this paper I will argue that North Korea's current nuclear posture is part of its strategy for economic renewal to survive."
SUBJECT(S):	Nuclear weapons--[North Korea] / Nuclear nonproliferation--[North Korea] / Nuclear research--[North Korea] / Economic conditions--[North Korea] / Foreign relations--[North Korea] / Foreign relations--[U.S.]--North Korea / Foreign relations--[North Korea]--U.S.
ADDED ENTRY:	Columbia University. East Asian Institute.

TITLE:	Rethinking about South Korea's security in face of North Korea's nuclear capability.
SOURCE:	Korea & world affairs, v. 18, summer 1994: 301-321.
AUTHOR:	Chee, Coung-Il.
LOCATION:	LRS94-9594 DS 920 E
NOTES:	"Tension over the North Korean nuclear issue is certainly mounting . . . . North Korea thus far has been successful in circumventing international pressure, including that of economic sanctions by the UN Security Council."
SUBJECT(S):	National security--[South Korea] / Nuclear weapons--[North Korea] / Inspection (Arms control)--[North Korea] / Nuclear nonproliferation / Foreign relations--[U.S.]--North Korea / Foreign relations--[North Korea]--U.S.

TITLE:	North Korea: just say no.
SOURCE:	America, v. 171, July 16, 1994: 12-15.
AUTHOR:	Bacevich, A. J.
LOCATION:	LRS94-9591 DS 930 E
NOTES:	"The real 'rising menace' that threatens the United States--though not it alone---is the continued legitimacy of nuclear weapons as instruments of politics."
SUBJECT(S):	Nuclear weapons--[North Korea] / Nuclear nonproliferation / Arms control / Foreign relations--[U.S.]--North Korea / Foreign relations--[North Korea]--U.S.

TITLE:	Nuclearization or denuclearization on the Korean Peninsula?
SOURCE:	Contemporary security policy, v. 15, Aug. 1994: 174-193.
AUTHOR:	Howlett, Darryl.
LOCATION:	LRS94-9535 DS 930 E
NOTES:	"The announcement by the Democratic People's Republic of Korea (DPRK) on 12 March 1993 that it was withdrawing from the Non-Proliferation Treaty (NPT) . . . . Indicated the possibility of a more disconcerting trend: the spread of nuclear weapons throughout the Northeast Asia region."
SUBJECT(S):	Nuclear weapons--[North Korea] / Nuclear weapons--[South Korea] / Arms control / Nuclear nonproliferation / Korean Peninsula

TITLE:	The world's most dangerous yard sale.
SOURCE:	Washington monthly, v. 26, Oct. 1994: 24-28.
AUTHOR:	Noah, Timothy.
LOCATION:	LRS94-9515 QC 170 U.S. A
NOTES:	Describes "how the U.S. government sold hardware and blueprints for a nuclear bomb to an Idaho used-car salesman who wants to sell them overseas."
SUBJECT(S):	Nuclear nonproliferation--[U.S.] / Nuclear facilities--[U.S.] / Nuclear weapons--[U.S.] / Privatization--[U.S.] / Defense policy--[U.S.]

TITLE:	". . . Carried by the wind out to sea" Ireland and the Isle of Man v. Sellafield: anatomy of a transboundary pollution dispute.
SOURCE:	Georgetown international environmental law review, v. 6, summer 1994: 639-681.
AUTHOR:	Hall, Terry.
LOCATION:	LRS94-9509 QC 170 For. Gen.

NOTES:	"Part I examines the origins and operations of Britain's nuclear reprocessing industry, its political and philosophical rationale, and its prospects. This Part also discusses the physical and psychological implications of exposure to radioactive substances--the core of the dispute --and outlines the attitudes and reactions of the Irish and Manx governments. Part II of the paper examines the various legal regimes available to the parties. The conclusions canvasses alternatives to these traditional avenues of redress, and concludes that current domestic and international legal systems are unequal to the task of effectively regulating the nuclear industry and its insidious emissions."
SUBJECT(S):	Nuclear nonproliferation--[Great Britain] / Transboundary pollution--[Great Britain] / Nuclear industry--[Great Britain] / Radiation safety--[Great Britain] / Radiation safety--[Ireland] / Nuclear industry--International cooperation / International environmental law / European Community

TITLE:	Russian policy and the Korean crisis.
SOURCE:	Carlisle Barracks, Pa., Strategic Studies Institute, U.S. Army War College, 1994. 26 p.
AUTHOR:	Blank, Stephen J.
LOCATION:	LRS94-9472 DK 350 C 4
NOTES:	"Dr. Blank relates Moscow's position on the issues of North Korean nuclearization to the broader domestic debate in Russia over security policy, in general, and Asian policy, in particular."
SUBJECT(S):	Nuclear weapons--[North Korea] / Arms control / Nuclear nonproliferation / Foreign relations--[Russia]--North Korea / Foreign relations--[North Korea]--Russia

TITLE:	Capping South Asia's nuclear weapons programs.
SOURCE:	Asian survey, v. 34, July 1994: 662-673.
AUTHOR:	Gordon, Sandy.
LOCATION:	LRS94-9413 DS 350
NOTES:	"The nuclear weapons programs of India and Pakistan are now well advanced. India, in particular, has a depth and breadth in its nuclear industries unmatched by any other threshold power. In these circumstances, it is doubtful whether attempts to force the South Asian threshold powers to roll back hard-won capabilities are any longer efficacious. If this is indeed the case, then major actors such as the United States will need to recognize the nuclear status quo in South Asia and begin to explore other means of preventing a spiraling nuclear competition from developing in the region."

Nuclear Proliferation: An Annotated Biography

SUBJECT(S): Nuclear weapons--[South Asia] / Nuclear nonproliferation--[South Asia]

TITLE: Nuclear deterrence in South Asia.
SOURCE: Asian survey, v. 34, July 1994: 647-661.
AUTHOR: Bhimaya, Kotera M.
LOCATION: LRS94-9412 DS 350

NOTES: "The U.S. and its allies argue that the deeply ingrained, traditional hostility between India and Pakistan accentuates the dangers inherent in a South Asian nuclear proliferation. In addition, they worry about the inadequacy of safeguards against accidents, the lack of circumspect behavior in decision-making; and whether command and control arrangements are sufficient to prevent a possible nuclear conflagration because of misperception, miscalculation, or both."

SUBJECT(S): Nuclear nonproliferation--[South Asia] / Nuclear weapons--[South Asia]

TITLE: International cooperation on nonproliferation export controls: prospects for the 1990s and beyond.
SOURCE: Ann Arbor, University of Michigan Press, 1994. 331 p.
LOCATION: LRS94-9389 LIMITED AVAILABILITY L SDI LoanJX1974.73.I584 1994 (LC no.)

NOTES: "This volume examines a wide range of countries representing a broad spectrum of political, economic, social, and security values: the United States, Russia, Germany, France, India, Brazil, Bulgaria, the former Czechoslovakia, Poland, and South Korea. The contributors look at the political, economic, bureaucratic, and military forces that are shaping each country's export control policies on strategic goods, technologies, and services. The case studies highlight the very different views held by these countries on the opportunities for, and constraints on, using export controls to further nonproliferation goals."

SUBJECT(S): Export controls / Nuclear nonproliferation
ADDED ENTRY: Bertsch, Gary K., Cupitt, Richard T., Elliott-Gower, Steven.

TITLE: Nuclear successor states of the Soviet Union.
SOURCE: Washington, Carnegie Endowment for International Peace ; Monterey, Calif., Monterey Institute of International Studies, 1994. 41 p.
LOCATION: LRS94-9085 F

NOTES: "Nuclear weapon and sensitive export status report number 1"

SUBJECT(S): Nuclear weapons--[CIS countries] / Arms control / Nuclear nonproliferation / Export controls / Nuclear exports--[CIS countries] / Belarus / Kazakhstan / Ukraine / Russia

TITLE: Arms control in the 1990s: proceedings of a workshop on chemical weapons, nuclear weapons and arms control in outer space held at Oakham House, Ryerson Polytechnic University on May 29th, 1993. Edited by Peter Brogden.
SOURCE: Ottawa, Canadian Centre for Global Security, 1994. 41 p. (Aurora papers no. 22)
LOCATION: LRS94-9084   FJK1974.A77 1994 (LC no.)

NOTES: Contents.--The Chemical Weapons Convention, by Walter Dorn. --Stability and arms control in outer space, by George Lindsey.--The NPT and CTBT--where are we getting to: a UN perspective, by Peggy Mason.

SUBJECT(S): Arms control--Conferences / Nuclear nonproliferation / Nuclear weapons tests / Chemical weapons / Space agreements / Canadian Centre for Global Security
ADDED ENTRY: Brogden, Peter.

TITLE: North Korea and the 'Madman' theory.
SOURCE: Security dialogue, v. 25, Sept. 1994: 307-316.
AUTHOR: Roy, Denny.
LOCATION: LRS94-8993   DS 930 E

NOTES: "This exceptionally careful handling of Pyongyang can be explained in large part by North Korea's reputation as a 'crazy' state . . . . I argue that North Korea's image as an irrational regime is largely a product of misunderstanding and propaganda."

SUBJECT(S): Nuclear weapons--[North Korea] / Politics and government--[North Korea] / Foreign relations--[U.S.]--North Korea / Foreign relations--[North Korea]--U.S. / Nuclear nonproliferation

TITLE: North Korea: enough carrots, time for the stick.
SOURCE: Comparative strategy, v. 13, 1994: 277-282.
AUTHOR: Bailey, Kathleen C.
LOCATION: LRS94-8351   DS 930 E

NOTES: "The United States should present Pyongyang with an ultimatum either to open up its nuclear facilities to special inspections or to face U.S. determination to inflict punishment with every means available short of war. By using its secret nuclear program for what amounts to extortion, North Korea is making a mockery of the nonproliferation regime."

SUBJECT(S): Nuclear weapons--[North Korea] / Inspection (Arms control)--[North Korea] / Foreign relations--[U.S.]--North Korea / Foreign relations--[North Korea]--U.S. / Nuclear nonproliferation

TITLE: Integration of the military and civilian nuclear fuel cycles in Russia.
SOURCE: Science & global security, v. 4, no. 3, 1994: 385-406.
AUTHOR: Bukharin, Oleg.
LOCATION: LRS94-7796   QC 170 For. CIS countries

NOTES: "This paper describes the close integration of the civil and military nuclear fuel cycles in Russia. Individual processing facilities, as well as the flow of nuclear material, are described as they existed in the 1980s and as they exist today. The end of the Cold War and the breakup of the Soviet Union weakened the ties between the two nuclear fuel cycles, but did not separate them. Separation of the military and civilian nuclear fuel cycles would facilitate Russia's integration into the world's nuclear fuel cycle and its participation in international nonproliferation regimes."

SUBJECT(S): Uranium--[Russia] / Nuclear fuels--[Russia] / Reactor fuel reprocessing--[Russia] / Nuclear energy policy--[Russia] / Nuclear nonproliferation--[Russia]

TITLE: For sale.
SOURCE: Newsweek, v. 124, Aug. 29, 1994: 30-32.
AUTHOR: Masland, Tom.
LOCATION: LRS94-7793   QC 170 Gen.

NOTES: "Deadly plutonium from Russia's vast nuclear network is turning up on the European market. Who is buying--and can they be stopped?"

SUBJECT(S): Plutonium--[Russia] / Nuclear nonproliferation--[Russia] / Nuclear fuels--[Russia] / Nuclear fuels--[Germany]

TITLE: The 1995 NPT Extension Conference.
SOURCE: Washington quarterly, v. 17, autumn 1994: 205-227.
AUTHOR: Lennon, Alexander T.
LOCATION: LRS94-7779   UA 17

NOTES: Contends that "nuclear nonproliferation as a whole is currently dependent upon the NPT itself [and] the multitude of issues for consideration at the conference--such as nuclear testing, security assurances, and strategic arms control--are merely symptomatic of a more fundamental concern among the NNWS: progress toward 'general and complete' nuclear disarmament."

SUBJECT(S): Nuclear nonproliferation

TITLE: Toward an international nuclear security policy.
SOURCE: Washington quarterly, v. 17, autumn 1994: 5-18.
AUTHOR: Quester, George H. Utgoff, Victor A.
LOCATION: LRS94-7772   US 17

NOTES: Article contends that "continued nuclear proliferation is going to change the nature of the world order in ways that will force the acceptance of strong and heretofore unpalatable countermeasures. Although the United States must take the initiative in defining and supporting the implementation of effective counters to proliferation, the definition and implementation of an effective policy must be an international undertaking."

SUBJECT(S): Nuclear nonproliferation / Arms control / Collective security

TITLE: Modeling decisionmaking of potential proliferators as part of developing counterproliferation strategies.
SOURCE: Santa Monica, Calif., Rand, 1994. 37 p.
AUTHOR: Arquilla, John. Davis, Paul K.
LOCATION: LRS94-7227   UA 17

NOTES: "This is a final report of a small exploratory project investigating new methods for understanding the reasoning and influencing the behavior of potential nuclear proliferators."

SUBJECT(S): Nuclear nonproliferation--Mathematical models / Nuclear weapons / Decision making
ADDED ENTRY: Rand Corporation.

TITLE: The domestic sources of regional regimes: the evolution of nuclear ambiguity in the Middle East.
SOURCE: International studies quarterly, v. 38, June 1994: 305-337.
AUTHOR: Solingen, Etel.
LOCATION: LRS94-7226   UV For. Middle East

NOTES: "The absence of a nuclear regime in the Middle East and the likely paths which may lead to one in the future. I identify four possible stylized outcomes: overt deterrence, regional 'opaqueness,' controlled proliferation, and a nuclear-weapons-free-zone."

SUBJECT(S): Nuclear weapons--[Middle East]--Future / Nuclear nonproliferation / Nuclear-weapon-free zones--[Middle East]

TITLE: North Korea and the challenge to the US-South Korean alliance.
SOURCE: Survival, v. 36, summer 1994: 78-91.
AUTHOR: Park, Sang Hoon.
LOCATION: LRS94-7185 DS 930 E

NOTES: "The purpose of this article is to analyse US and South Koran approaches to the North Korean nuclear issue, focusing on the differences in perceptions and policy priorities."

SUBJECT(S): Nuclear weapons--[North Korea] / Nuclear nonproliferation / Inspection (Arms control)--[North Korea] / Foreign relations--[U.S.]--South Korea / Foreign relations--[South Korea]--U.S. / Foreign relations--[South Korea]--North Korea / Foreign relations--[North Korea]--South Korea

TITLE: Proliferation and nonproliferation in Ukraine: implications for European and U.S. security.
SOURCE: Carlisle Barracks, Pa., Strategic Studies Institute, U.S. Army War College, 1994. 37 p.
AUTHOR: Blank, Stephen J.
LOCATION: LRS94-7016 DK 508

NOTES: "This study explores the background, terms, and aftermath of the January 1994 tripartite agreement among Russia, Ukraine, and the United States concerning the removal of nuclear missiles located in Ukraine after the fall of the Soviet Union. This chapter in international security raises difficult issues for the United States concerning regional security and nuclear proliferation."

SUBJECT(S): Nuclear weapons--[Ukraine] / Nuclear nonproliferation / National security--[Europe] / National security--[U.S.] / International relations / Arms control
ADDED ENTRY: Army War College (U.S.). Strategic Studies Institute.

TITLE: Fixing Jimmy Carter's mistakes: regaining the initiative against North Korea.
SOURCE: Washington, Heritage Foundation, Asian Studies Center, 1994. 9 p. (Backgrounder no. 131)
AUTHOR: Fisher, Richard D., Jr.
LOCATION: LRS94-7010 JK 516 A 9

NOTES: "America and its Asian allies must regain the initiative in convincing and, if necessary, compelling North Korea to give up its nuclear weapons program."

SUBJECT(S): Nuclear weapons--[North Korea] / Nuclear nonproliferation /

ADDED ENTRY:	Foreign relations--[U.S.]--North Korea / Foreign relations--[North Korea]--U.S. / Carter, Jimmy. Asian Studies Center (Heritage Foundation).

TITLE:	Nuclear threats from small states.
SOURCE:	Carlisle Barracks, Pa., Strategic Studies Institute, U.S. Army War College, 1994. 22 p.
AUTHOR:	Kahan, Jerome H.
LOCATION:	LRS94-6691 UC 650 A
NOTES:	"What are the policy implications regarding proliferation and counterproliferation of nuclear weapons among Third World States? .... Mr. Jerome Kahan examines the likelihood that one or more of these countries will use nuclear weapons before the year 2000."
SUBJECT(S):	Nuclear weapons / Small states / Military policy--[U.S.] / Third World / Nuclear nonproliferation / Arms control

TITLE:	Syndicated columnists.
SOURCE:	Washington post, June 5, 1994: C7; June 9: A27.
AUTHOR:	Will, George F.
LOCATION:	LRS94-6283 PN U.S. B Will
NOTES:	Contents.--Young men who dared.--. . . Nor time to waste on Korea.
SUBJECT(S):	World War II / Veterans / Nuclear weapons (North Korea) / Nuclear nonproliferation / Foreign relations--[U.S.]--North Korea / Foreign relations--[North Korea]--U.S.

TITLE:	A long-term plan of R & D and use of atomic energy [Genshiryoku no kenkyu, kaihatsu oyobi riyo ni kansuru chokikeikaku]
SOURCE:	Tokyo, Science and Technology Agency, 1994. 87 p.
AUTHOR:	Japan. Atomic Energy Committee.
LOCATION:	LRS94-6138 JDC
NOTES:	Considers development and use of atomic energy in Japan effective until 2010. Shows Japan's fundamental concepts on what its atomic energy development and use bring to a global society, why Japan employs nuclear fuel recycling policy. Japan is thinking about non-proliferation of nuclear weapons and its peaceful use and how Japan should proceed with peaceful use of atomic energy such as plutonium. This long-term plan is a result of deliberation by specialists and laymen from various fields. Contents.--Global society in the 21st century and role of atomic energy.--What Japan's development and use of atomic energy should be.--Japan's future plan for development and use of atomic energy including securing of safety.

technical development of nuclear fuel recycling, international cooperation.
Text in Japanese; citation title and abstract translated by the Japan Documentation Center.

SUBJECT(S): Nuclear energy--[Japan] / Nuclear energy policy--[Japan] / Nuclear nonproliferation--[Japan] / Peaceful uses of nuclear energy--[Japan]
ADDED ENTRY: Japan. Genshiryoku Iinkai.

TITLE: The security situation on the Korean Peninsula. Joint hearing before the Subcommittees on International Security, International Organizations and Human Rights and Asia and the Pacific of the Committee on Foreign Affairs, House of Representatives, 103d Congress, 2nd session. Feb. 24, 1994.
SOURCE: Washington, G.P.O., 1994. 80 p.
AUTHOR: U.S. Congress. House. Committee on Foreign Affairs. Subcommittee on International Security, International Organizations and Human Rights.
LOCATION: LRS94-6064   CONGRESSIONAL PUBLICATION   RBC 6780

NOTES: "Threat of nuclear proliferation in North Korea."

SUBJECT(S): Nuclear weapons--[North Korea] / Nuclear nonproliferation / Inspection (Arms control)--[North Korea] / Foreign relations--[U.S.]--North Korea / Foreign relations--[North Korea]--U.S. / National security--[South Korea] / Korean Peninsula
ADDED ENTRY: U.S. Congress. House. Committee on Foreign Affairs. Subcommittee on Asia and the Pacific.

TITLE: Russia: Delusions v. conversion.
SOURCE: Bulletin of the atomic scientists, v. 50, July-Aug. 1994: 11-13.
AUTHOR: Khripunov, Igor.
LOCATION: LRS94-6031   DK 350 E

NOTES: Contends that "Russian defense conversion is a gloomy story, punctuated by only a few isolated successes."

SUBJECT(S): Defense industries--[Russia] / Nuclear nonproliferation--[Russia] / Conversion of industries--[Russia]

TITLE: Table of contents page.
SOURCE: Bulletin of the atomic scientists, v. 50, July-Aug. 1994.
LOCATION: LRS94-6030   Optical Disk

NOTES: Copy of the contents page of the journal annotated with document identification numbers (LRS numbers) to facilitate the retrieval of individual articles from the CRS optical disk system. To view the table of contents on the SCORPIO system, browse the title of the journal, select the set and display the most recent citations.

SUBJECT(S): Nuclear nonproliferation

TITLE: Who's threatening who?
SOURCE: Technology review, v. 97, July 1994: 72-73.
AUTHOR: Hart-Landsberg, Martin.
LOCATION: LRS94-5819   DS 930 E

NOTES: "To resolve the nuclear faceoff with North Korea, the United States will have to rethink its singleminded and hostile approach."

SUBJECT(S): Nuclear weapons--[North Korea] / Nuclear nonproliferation / Inspection (Arms control)--[North Korea]

TITLE: The North Korean nuclear program: What is to be done?
SOURCE: Santa Monica, Calif., Rand Corporation, 1994. 27 p. (MR-434-A)
AUTHOR: Wendt, James C.
LOCATION: LRS94-5728   DS 930 E

SUBJECT(S): Nuclear weapons--[North Korea] / Nuclear nonproliferation / Foreign relations--[U.S.]--North Korea / Foreign relations--[North Korea]--U.S
ADDED ENTRY: Rand Corporation.

TITLE: [Non-proliferation]
SOURCE: Orbis, v. 38, summer 1994: 409-456.
AUTHOR: Steinberg, Gerald M.   Krasno, Jean.   Chellaney, Brahma.
LOCATION: LRS94-5530   UA 17

NOTES: Series of three articles examines nonproliferation issues. Contents.--Time for regional approaches?, by Gerald M. Steinberg.--Brazil's secret nuclear program, by Jean Krasno.--An Indian critique of U.S. export controls, by Brahma Chellaney.

SUBJECT(S): Nuclear nonproliferation

TITLE: Nuclear proliferation: U.S. aims and India's response.
SOURCE: Studies in conflict and terrorism, v. 17, 1994: 165-180.
AUTHOR: Nair, Vijai K.
LOCATION: LRS94-5209   UV For. India

NOTES: "The NPT is a contentious issues between the world's two largest democracies . . . . The United States and India need to review their stands . . . and work to lessen the gap between their respective policies without impinging on the other's legitimate national interests."

SUBJECT(S): Nuclear nonproliferation / Nuclear weapons--[India] / Foreign relations--[U.S.]--India / Foreign relations--[India]--U.S.

TITLE: Bridging the nuclear impasse on the Korean peninsula.
SOURCE: Journal of international affairs, v. 4, spring 1994: 62-66.
AUTHOR: Boller, Scott C.
LOCATION: LRS94-5198 DS 930 E

NOTES: "Some of the motivations behind North Korea's actions."

SUBJECT(S): Nuclear weapons--[North Korea] / Nuclear nonproliferation / Inspection (Arms control)--[North Korea]

TITLE: North Korea and the challenge to the US-South Korean alliance.
SOURCE: Survival, v. 36, summer 1994: 78-91.
AUTHOR: Park, Sang Hoon.
LOCATION: LRS94-5194 DS 930 E

NOTES: "The purpose of this article is to analyse US and South Korean approaches to the North Korean nuclear issue, focusing on the differences in perceptions and policy priorities."

SUBJECT(S): Nuclear weapons--[North Korea] / Nuclear nonproliferation / Foreign relations--[U.S.]--South Korea / Foreign relations--[South Korea]--U.S. / Inspection (Arms control)--[North Korea]

TITLE: The dangerous myths of START II.
SOURCE: Parameters, v. 24, spring 1994: 78-87.
AUTHOR: Boldrick, Michael R.
LOCATION: LRS94-5189 UA 17

NOTES: Addresses six myths of treaty supporters.

SUBJECT(S): Nuclear weapons / Nuclear weapons--[Russia] / Nuclear nonproliferation / National security--[U.S.] / Strategic Arms Reduction Talks

TITLE: The Continuing North Korean nuclear crisis.
SOURCE: Arms control today, v. 24, June 1994: 18-22.
LOCATION: LRS94-5127 DS 930 E

NOTES: "The confrontation between North Korea and the International Atomic Energy Agency (IAEA) over the right to fully inspect Pyongyang's nuclear facilities has continued for over a year now. Recent diplomatic activities and statements have prompted a new look at the ongoing crisis. On May 5 the Arms Control Association (ACA) held a news conference to examine North Korea's refusal to abide by its commitments under the nuclear Non-Proliferation (NPT) and to assess recent developments. Panelists included Spurgeon M. Kenny, Jr., . . . Jon Wolfsthal; William Dircks, . . . and William Taylor."

SUBJECT(S): Nuclear research--[North Korea] / Nuclear weapons--[North Korea] / Nuclear facilities--[North Korea] / Nuclear nonproliferation--[North Korea] / Inspection (Arms control)

TITLE: Nuclear nonproliferation: licensing procedures for dual-use exports need strengthening; testimony before the Senate Committee on Governmental Affairs. May 17, 1994.
SOURCE: Washington, G.A.O., 1994. 12 p.
AUTHOR: U.S. General Accounting Office.
LOCATION: LRS94-4989 AVAILABLE FROM ISSUING AGENCY Optical Disk

NOTES: "GAO/T-NSIAD-94-163"
"Statement of Joseph E. Kelley, Director-in-Charge, International Affairs Issues, National Security and International Affairs Division."

SUBJECT(S): Nuclear nonproliferation / Export controls--[U.S.]

TITLE: Defusing North Korea's nuclear threat.
SOURCE: Washington, Heritage Foundation, 1994. 5 p. (Backgrounder no. 224)
AUTHOR: Plunk, Daryl M.
LOCATION: LRS94-4886   DS 930 C 2

NOTES: "The White House has been pursuing an unnecessarily protracted negotiation strategy and has left unclear its terms for a final resolution of the dispute . . . . President Clinton should communicate to the North, and to the American people, his intention to establish first-ever diplomatic ties with Pyongyang and end the U.S. trade embargo against North Korea once the nuclear issue has been resolved."

SUBJECT(S): Foreign relations--[U.S.]--North Korea / Foreign relations--[North Korea]--U.S. / Nuclear weapons--[North Korea] / Nuclear nonproliferation / Inspection (Arms control)--[North Korea]
ADDED ENTRY: Heritage Foundation.

TITLE:	The wild East.
SOURCE:	Atlantic monthly, v. 273, June 1994: 61-86 passim (15 p.)
AUTHOR:	Hersh, Seymour M.
LOCATION:	LRS94-4875 HV 6251 E
NOTES:	"Intelligence reports from a variety of sources suggest that it may be only a matter of time before the Russian mafia, which now controls most of Russia's economy, gets its hands on nuclear materials--if it has not done so already."
SUBJECT(S):	Organized crime--[Russia] / Corporate corruption--[Russia] / Nuclear terrorism--[Russia]--Scenarios / Nuclear nonproliferation--Scenarios

TITLE:	Export controls and nonproliferation policy.
SOURCE:	Washington, Office of Technology Assessment, for sale by the Supt. of Docs., G. P.O., 1994. 82 p.
LOCATION:	LRS94-4836 AVAILABLE FROM ISSUING AGENCY HF 1455
NOTES:	"Export controls on dual-use goods, technology, and software will continue to be one useful tool in U.S. efforts to stem the proliferation of weapons of mass destruction and missiles that can deliver them." Partial contents.--Current United States nonproliferation export controls.--Weighing benefits and costs.--Options for enhancing export controls.--Reducing the burdens on industry.
SUBJECT(S):	Export controls--[U.S.] / Nuclear nonproliferation / Arms control
ADDED ENTRY:	U.S. Congress. Office of Technology Assessment.

TITLE:	Extended conventional deterrence: in from the cold and out of the nuclear fire.
SOURCE:	Washington quarterly, v. 17, summer 1994: 203-233.
LOCATION:	LRS94-4798 UA 17
NOTES:	"The purpose of this research survey is to determine the outline of U.S. post-cold war thinking on deterrence, while focusing on extended conventional deterrence."
SUBJECT(S):	Deterrence / Nuclear nonproliferation / Arms control
ADDED ENTRY:	Allan, Charles T.

TITLE:	Facing the emerging reality of regional nuclear adversaries.
SOURCE:	Washington quarterly, v. 17, summer 1994: 41-71.
AUTHOR:	Millot, Marc Dean.
LOCATION:	LRS94-4789 UA 17

NOTES: Examines necessary steps to develop a new military strategy.

SUBJECT(S): Arms control / Nuclear nonproliferation / Military policy--[U.S.] / Military strategy--[U.S.]

TITLE: On dealing with the prospect on nuclear chaos.
SOURCE: Washington quarterly, v. 17, summer 1994: 19-339.
AUTHOR: Molander, Roger C. Wilson, Peter A.
LOCATION: LRS94-4788   UA 17

NOTES: Examines "the possible role of nuclear weapons in the international security environments that may unfold in the course of the next generation--say, the next 20 to 25 years."

SUBJECT(S): Nuclear weapons / Arms control / Nuclear nonproliferation

TITLE: The last nuclear summit?
SOURCE: Washington quarterly, v. 17, summer 1994: 5-15.
AUTHOR: Reiss, Mitchell.
LOCATION: LRS94-4787   UA 17

NOTES: "Central to past success in the nuclear sphere has been the Nuclear Non-Proliferation Treaty (NPT) and the associated safeguards system applied by the International Atomic Energy Agency (IAEA). Multiplying political instabilities and the dissemination of advanced technologies make this global regime even more important in the post-cold world." Discusses issues concerning its extension.

SUBJECT(S): Arms control agreements / Nuclear nonproliferation / Nuclear weapons

TITLE: Secrets of the Soviet nuclear complex.
SOURCE: IEEE spectrum, v. 31, May 1994: 32-38.
AUTHOR: Kalinin, Alexander V.
LOCATION: LRS94-4503   QC 170 For. CIS countries

NOTES: Profiles and describes the nuclear weapons, nuclear power plants and nuclear research facilities of the former Soviet Union.

SUBJECT(S): Nuclear energy policy--[CIS countries] / Nuclear security measures--[CIS countries] / Nuclear facilities--[CIS countries]--Maps / Nuclear weapons--[CIS countries] / Nuclear power plants--[CIS countries] / Nuclear nonproliferation--[CIS countries]

TITLE:	Denuclearization in Argentina and Brazil.
SOURCE:	Arms control today, Mar. 1994: 10-14.
AUTHOR:	Goldemberg, Jose. Feiveson, Harold A.
LOCATION:	LRS94-4403 UV For. Latin America
NOTES:	"The decisions by Argentina and Brazil to forgo the production of nuclear weapons is closely linked to the return of democratic rule in both countries after decades of military governments."
SUBJECT(S):	Nuclear nonproliferation--[Argentina] / Arms control--[Latin-America] / Nuclear nonproliferation--[Brazil]

TITLE:	Nuclear nonproliferation: export licensing procedures for dual-use items need to be strengthened; report to the chairman, Committee on Governmental Affairs, U.S. Senate. Apr. 26, 1994.
SOURCE:	Washington, G.A.O., 1994. 69 p.
AUTHOR:	U.S. General Accounting Office.
LOCATION:	LRS94-4349 AVAILABLE FROM ISSUING AGENCY Optical Disk
NOTES:	"GAO/NSIAD-94-119, B-256585" "Export licensing procedures for dual-use nuclear items . . . . How well the U.S. government is implementing policies and procedures to prevent exports that pose a proliferation risk."
SUBJECT(S):	Nuclear nonproliferation / Arms control / Export controls--[U.S.]

TITLE:	Will Ukraine go nuclear?
SOURCE:	Mediterranean quarterly, v. 5, winter 1994: 21-35.
AUTHOR:	Moshes, Arkady.
LOCATION:	LRS94-4114 DK 508
NOTES:	"In spite of the agreements, is this regional military superpower coming closer and closer to nuclear status? . . . . In order to have a better understanding of the present, we must review the evolution of the Ukrainian approach to nuclear weapons problems from 1990 to 1993."
SUBJECT(S):	Nuclear weapons--[Ukraine] / Nuclear nonproliferation

TITLE:	The perils of proliferation: organization theory, deterrence theory, and the spread of nuclear weapons.
SOURCE:	International security, v. 18, spring 1994: 66-107.
AUTHOR:	Sagan, Scott D.
LOCATION:	LRS94-3862 UA 17
NOTES:	"What are the likely effects of the spread of nuclear weapons?"

SUBJECT(S):	Nuclear weapons / Nuclear nonproliferation / Deterrence

TITLE:	Nuclear nonproliferation: concerns with U.S. delays in accepting foreign research reactors' spent fuel; report to congressional requesters. Mar. 25, 1994.
SOURCE:	Washington, G.A.O., 1994. 29 p.
AUTHOR:	U.S. General Accounting Office.
LOCATION:	LRS94-3818 AVAILABLE FROM ISSUING AGENCY Optical Disk
NOTES:	"GAO/RCED-94-119, B-256337" "A key nonproliferation goal of the United States is to discourage the use of highly enriched uranium fuel (HEU), a material that can be used to make nuclear bombs .... U.S. officials question the safety of spent (used) HEU fuel left in interim storage at reactor sites throughout the world and, for security reasons, would prefer that this spent fuel be consolidated and permanently stored in the United States."
SUBJECT(S):	Uranium enrichment--[Foreign]--Safety measures / Nuclear nonproliferation / Arms control

TITLE:	North Korea in 1993: in the eye of the storm.
SOURCE:	Asian survey, v. 34, Jan. 1994: 10-18.
AUTHOR:	Merrill, John.
LOCATION:	LRS94-3667 DS 930 E
NOTES:	"The Democratic People's Republic of Korea (DPRK) was the focus of international attention in 1993 because of the controversy over its nuclear weapons program."
SUBJECT(S):	Nuclear research--[North Korea] / Nuclear facilities--[North Korea] / Nuclear nonproliferation / Inspection (Arms control)--[North Korea] / Foreign relations--[North Korea]

TITLE:	North Korea's nuclear challenge.
SOURCE:	Korea and world affairs, v. 18, spring 1994: 23-41.
AUTHOR:	Rubinstein, Alvin Z.
LOCATION:	LRS94-3586 DS 930 E
NOTES:	"North Korea's apparent determination to go nuclear confronts South Korea, Japan, China, the United States, and the IAEA with a number of far-reaching policy-relevant questions .... Each of the parties affected has a different assessment of and suggested approach to the situation. Their leaderships do not agree on the nature of the North Korean threat, much less on how to cope with it."
SUBJECT(S):	Nuclear research--[North Korea] / Nuclear facilities--[North Korea] / Nuclear weapons--[North Korea] / Nuclear

nonproliferation / Inspection (Arms control)

TITLE:	Nuclear nonproliferation: concerns with U.S. delays in accepting foreign research reactors' spent fuel; report to congressional requesters. Mar. 25, 1994.
SOURCE:	Washington, G.A.O., 1994. 29 p.
AUTHOR:	U.S. General Accounting Office.
LOCATION:	LRS94-3498 AVAILABLE FROM ISSUING AGENCY Optical Disk
NOTES:	"GAO/RCED-94-119, B-256337" "Key nonproliferation goal of the United States is to discourage the use of highly enriched uranium fuel (HEU), a material that can be used to make nuclear bombs. U.S. officials would prefer that this spent fuel be consolidated and permanently stored in the United States."
SUBJECT(S):	Uranium enrichment--[Foreign] / Nuclear fuels--[Foreign] / Nuclear nonproliferation / Arms control

TITLE:	A bomb waiting to explode.
SOURCE:	New scientist, v. 14, Feb. 26, 1994: 14-15.
AUTHOR:	Kiernan, Vincent.
LOCATION:	LRS94-3428 QC 170 Gen.
NOTES:	"The world's stockpile of plutonium could reach 1700 tonnes by the end of the century. You need 5 kilograms to make a bomb."
SUBJECT(S):	Plutonium / Nuclear nonproliferation / Nuclear terrorism

TITLE:	Dangerous surplus.
SOURCE:	Bulletin of the atomic scientists, v. 50, May-June 1994: 39-41.
AUTHOR:	Holdren, John P.
LOCATION:	LRS94-3368 UC 650 A
NOTES:	As a result of arms-control reduction, weapons-grade plutonium will become surplus in the United States and Russia. "It is crucial that this surplus weapons plutonium be managed in a way that minimizes the danger that it will be re-used for weapons by the initial possessor nation, another nation, or a subnational group; strengthens national and international institutions and incentives for control and reduction of nuclear weapons; does not lead to increased accessibility of civilian plutonium for weapons use; and meets reasonable standards for safety, health, the environment, and cost."
SUBJECT(S):	Plutonium / Nuclear nonproliferation / Nuclear weapons / International control of nuclear power / Nuclear security measures

TITLE:	Table of contents page.
SOURCE:	Bulletin of the atomic scientists, v. 50, May-June 1994.
LOCATION:	LRS94-3365　　　　　　　　　　　　　　　　Optical Disk
NOTES:	Copy of the contents page of the journal annotated with document identification numbers (LRS numbers) to facilitate the retrieval of individual articles from the CRS optical disk system. To view the table of contents on the SCORPIO system, browse the title of the journal, select the set and display the most recent citations.
SUBJECT(S):	Nuclear nonproliferation

TITLE:	Safeguarding the ingredients for making nuclear weapons.
SOURCE:	Issues in science and technology, v. 10, spring 1994: 67-73.
AUTHOR:	Panofsky, Wolfgang K. H.
LOCATION:	LRS94-3181　　　　　　　　　　　　　　　　　　UC 650 A
NOTES:	"With the end of the Cold War, concerns about the proliferation of nuclear arms have come to seem far more important than preventing or preparing for direct nuclear conflict. Controlling proliferation will necessitate dealing with technical, institutional, and political difficulty is limiting access to fissionable material--the plutonium or highly enriched uranium (HEU) that can be used to make a nuclear weapon."
SUBJECT(S):	Nuclear nonproliferation--[U.S.] / Nuclear nonproliferation--[Russia] / Nuclear weapons / Nuclear security measures--[U.S.] / Nuclear security measures--[Russia]

TITLE:	Fighting proliferation with intelligence.
SOURCE:	Orbis, v. 38, spring 1994: 245-260.
AUTHOR:	Sokolski, Henry.
LOCATION:	LRS94-2825　　　　　　　　　　　　　　　　　　UV Gen.
NOTES:	"As difficult as defining proliferation may be, the intelligence and policy communities should make the effort."
SUBJECT(S):	Intelligence activities / Nuclear nonproliferation

TITLE:	Tensions on the Korean Peninsula. Hearing, 103d Congress, 1st session. Nov. 3, 1993.
SOURCE:	Washington, G.P.O., 1994. 59 p.
AUTHOR:	U.S. Congress. House. Committee on Foreign Affairs. Subcommittee on Asia and the Pacific.
LOCATION:	LRS94-2758　　CONGRESSIONAL PUBLICATION　　　　RBC 6367

NOTES:	"The seriousness with which the international community views North Korea's apparent determination to proceed with the development of nuclear weapons."
SUBJECT(S):	Nuclear research--[North Korea] / Nuclear weapons--[North Korea] / Foreign relations--[U.S.]--North Korea / Foreign relations--[North Korea]--U.S., / Inspection (Arms control)--[North Korea] / Nuclear nonproliferation

TITLE:	North Korea: a potential time bomb.	
SOURCE:	Jane's intelligence review, no. 2, 1994, suppl.: 1-24.	
LOCATION:	LRS94-2523	DS 930 E
SUBJECT(S):	Nuclear weapons--[North Korea] / Nuclear research--[North Korea] / Nuclear facilities--[North Korea] / Inspection (Arms control)--[North Korea] / Nuclear nonproliferation / Chemical weapons--[North Korea] / Biological weapons--[North Korea] / Ballistic missiles--[North Korea] / Army--[North Korea]	

TITLE:	Management and disposition of excess weapons plutonium.	
SOURCE:	Arms control today, v. 24, Mar. 1994: 27-31.	
LOCATION:	LRS94-2455	UC 650 A
NOTES:	Present excerpts from a 36-page executive summary of a National Academy of Science report that recommended "a comprehensive approach to the handling of the large stocks of weapons plutonium no longer needed with the end of the Cold War. The study, entitled 'Management and Disposition of Excess Eapons Plutonium,' presents detailed recommendations on a reciprocal U.S- Russian plutonium regime, which would include: declarations on total inventories of weapons and fissile materials, monitored dismantlement of weapons, safeguarded interim storage of materials, and long-term disposal of excess plutonium either by vitrification into. large logs with high-level waste or by use in existing reactors without future reprocessing."	
SUBJECT(S):	Plutonium / Radioactive waste disposal / Nuclear weapons / Nuclear security measures / Nuclear nonproliferation / International control of nuclear power	

TITLE:	North Korea's nuclear threat challenges the world and tests America's resolve.	
SOURCE:	Washington, Heritage Foundation, 1994. 12 p. (Asian Studies Center backgrounder no. 129)	
AUTHOR:	Fisher, Richard D., Jr.	
LOCATION:	LRS94-2194	DS 930 E

NOTES: "The most critical foreign policy challenge now facing President Bill Clinton is how to terminate North Korea's nuclear weapons program .... If diplomacy fails, the options open to the United States and South Korea are grim. If Washington and Seoul seek economic sanctions to compel nuclear inspections, they risk war."

SUBJECT(S): Nuclear weapons--[North Korea] / Nuclear nonproliferation / Inspection (Arms control)--[North Korea] / Foreign relations--[U.S.]--North Korea / Foreign relations--[North Korea]--U.S. / Arms control

ADDED ENTRY: Heritage Foundation.

TITLE: Science after the Cold War: a special report.
SOURCE: Science, v. 263, Feb. 4, 1994: 619-631.
LOCATION: LRS94-2114　　　　　　　　　　　　　　　　　　Q 125 U.S. C

NOTES: Contents.--Researchers shift the ashes of SDI.--Ex-defense scientists come in from the cold.--The Defense Department declassifies the earth-slowly.--Nonproliferation boom gives a lift to the national labs.--Plutonium disposal: no easy way to shackle the nuclear demon.

SUBJECT(S): Science policy--[U.S.] / Military research--[U.S.] / Nuclear nonproliferation--[U.S.] / Research and development facilities--[U.S.] / Plutonium / Radioactive waste disposal --[U.S.]

TITLE: No-first-use and nonproliferation: redefining extended deterrence.
SOURCE: Washington quarterly, v. 17, spring 1994: 103-114.
AUTHOR: Quester, George H. Utgoff, Victor A.
LOCATION: LRS94-1441　　　　　　　　　　　　　　　　　　UA 17

NOTES: "This article offers a critique of the conventional wisdoms about no-first-use and nonproliferation. It begins with a review of the no-first-use in U.S. nuclear doctrine and both the opportunities and challenges of the new era. It assesses the type of balance that must be struck between conventional and nuclear means of deterring renegade, nuclear-armed states."

SUBJECT(S): Nuclear nonproliferation

TITLE: Life after proliferation.
SOURCE: Foreign affairs, v. 73, Mar.-Apr. 1994: 8-13.
AUTHOR: Carpenter, Ted Galen.
LOCATION: LRS94-1183　　　　　　　　　　　　　　　　　　UA 17

NOTES: "It is time for U.S. leaders to reassess Cold War policies on nonproliferation, security commitments and extended deterrence and to adapt them to changed international circumstances. These commitments may once have made sense, given the need to thwart the Soviet Union's expansionist agenda. But they are highly dubious in the absence of the superpower rivalry. They now threaten to embroil the United States in regional conflicts where nuclear weapons have already proliferated or will inevitably proliferate soon."

SUBJECT(S): Nuclear nonproliferation

TITLE: Arming for peace.
SOURCE: Bulletin of the atomic scientists, v. 50, Mar.-Apr. 1994: 38-43.
AUTHOR: Tsipis, Kosta. Morrison, Philip.
LOCATION: LRS94-1040   UC 650 A

NOTES: "Under a U.N.-sponsored nuclear umbrella, nations could devote resources to improving the quality of life rather than the quantity of weapons."

SUBJECT(S): Nuclear nonproliferation

TITLE: NATO: skittish on counterproliferation.
SOURCE: Bulletin of the atomic scientists, v. 50, Mar.-Apr. 1994: 12-13.
AUTHOR: Goldring, Natalie J.
LOCATION: LRS94-1036   JX 4005 C

NOTES: "The United States wants its allies to join in a 'counterproliferation' effort. However, NATO members, who were skeptical of Star Wars, are not sure they want take part--militarily or financially--in a U.S. plan to develop ballistic missile defenses against nuclear weapons."

SUBJECT(S): Advanced weapons--[NATO countries] / NATO military forces / Nuclear nonproliferation--[NATO countries]

TITLE: Nuclear non-proliferation in the post-Cold War era.
SOURCE: International affairs (London), v. 70, Jan. 1994: 17-39.
AUTHOR: Simpson, John.
LOCATION: LRS94-650   UA 17

NOTES: "New proliferation risks, and new challenges to the international consensus underpinning the non-proliferation regime, are identified and discussed. It is argued that, given this context, the 1995 conference could be a watershed for the regime."

SUBJECT(S): Nuclear nonproliferation

TITLE:	Back to the future: toward a post-nuclear ethic--the new logic of nonproliferation.
SOURCE:	Washington, Progressive Foundation, 1994. 54 p.
AUTHOR:	Manning, Robert A.
LOCATION:	LRS94-385 UC 650 A
SUBJECT(S):	Nuclear nonproliferation--Future / Nuclear weapons--Future / Arms control--Future / Nuclear weapons tests--Future / Military policy--[U.S.]--Future
ADDED ENTRY:	Progressive Foundation.

TITLE:	U.S. and Russian military-technical policy: conference.
SOURCE:	Comparative strategy, v. 13, no. 1, 1994: whole issue (156 p.)
LOCATION:	LRS94-384 LIMITED AVAILABILITY L SDI Loan
NOTES:	Partial contents.--U.S. and Russian military-technical policy.--Proliferation, deterrence, stability and missile defense.
SUBJECT(S):	Military policy--[Russian Republic] / Defense industries--[Russian Republic] / Military relations--[U.S.]--Russian Republic / Military relations--[Russian Republic]--U.S. / Defense industries--[U.S.] / Conversion of industries--[Russian Republic] / Deterrence / Ballistic missile defenses / Nuclear nonproliferation

TITLE:	Proliferation: bronze medal technology is enough.
SOURCE:	Orbis, v. 38, winter 1994: 67-82.
AUTHOR:	Zimmerman, Peter D.
LOCATION:	LRS94-153 UV Gen.
NOTES:	"The industrialized countries of the world normally view long-range guided missiles and nuclear weapons as developments requiring the best of First World technology to design and produce. In fact, neither missiles nor the weapons of mass destruction that go atop them require 1990s or even 1970s technology. Both types of weapons were fully mature before 1960; the long-range rocket entered military service in 1944, and the nuclear weapon in 1945. Understanding the implications of this history is critical to formulating and implementing successful antiproliferation policies for the 1990s."
SUBJECT(S):	Weapons of mass destruction / Advanced weapons / Nuclear nonproliferation

TITLE:	Rethink the nuclear threat.
SOURCE:	Orbis, v. 38, winter 1994: 99-108.
AUTHOR:	Powers, John R. Muckerman, Joseph E.
LOCATION:	LRS94-38 UC 650 A
NOTES:	"Contrary to the conventional wisdom, the danger of the use of nuclear weapons is greater now than at any time since Hiroshima. The increased risks are due mainly to political instabilities throughout the world fueled by ethnic conflicts, militant fundamentalism, and terrorism; fundamental economic problems afflicting the nuclear republics of the former Soviet Union as well as many other nations; and unbridled proliferation of nuclear weapons and their means of delivery."
SUBJECT(S):	Nuclear weapons / Nuclear nonproliferation

Subject Index

13th World Congress of the International Political Science Association, 24
1993-North Korea, 218
98th Congress-1st Session, 83

A

Advanced Weapons-(NATO countries), 223
Africa, 124, 125
America and the World of the 21st Century, 35
American Atom, 2
American Military Assistance, 157
American Nuclear Society, 65
Anarchy, 138
Antimissile Missiles, 134
Antinuclear weapons movement, 158
Argentina, 159, 217
Arms Control Agreement, 123, 126, 132, 141, 143, 149, 151
Arms Control and Disarmament Agency, 84
Arms Control and Disarmament Agency-Appropriations and Expenditures, 101
Arms Control and disarmament division, 153
Arms Control Verification (North Korea), 153
Arms Control Verification, 53, 86, 114, 119, 120
Arms Control, 3, 10, 15, 30, 52, 66, 73, 78, 93, 139, 142, 146, 148, 150, 154, 157, 162, 163, 218
Arms Control-East Asia, 61
Arms Control-Europe, 61
Arms Control-Former Soviet Republics, 105
Arms Control-Russia (Federation), 102
Arms Control-United States, 108
Arms Control Verification, 119, 120
Arms Export Licensing, 107
Arms Production, 9
Arms Proliferation, 3, 4
Arms Race, 5, 49, 58, 73, 78, 95, 105, 133
Arms Race-East Asia, 136
Arms Sales-Russia, 143
Arms Transfers, 92
Arms Transfers-China, 88
Arms Transfers-Pakistan, 96
Arroyo Center, 114
Asia and the Pacific, 86, 176
Aspen Strategy Group, 38
Atomic Bomb, 127, 131, 196
Atomic Bomb-Materials, marketing, prevention, 101
Australia-Foreign Relations, 114

B

Bahais-Civil Rights-Iran, 85
Ballistic Missile Defenses, 76, 90, 97, 102, 133, 134, 156, 179,
Ballistic Missiles-North Korea, 149
Belarus, 201, 206
Biological Arms control, 40, 50, 60, 99, 104, 130, 133, 160
Blocking the Spread of Nuclear Weapons, 9, 10
Brain Drain-Russia, 60, 115, 179
Brazil, 124, 159, 217

Bush, George President (1989-1993), 112

C

Canada Dept of Foreign Affairs and international Trade, 153
Canadian Centre for Arms Control and Disarmament, 114
Canadian Centre for Global Security, 57, 206
Canberra Commission on the Elimination of Nuclear Weapons, 11
Carnegie Commission, 13, 14
Castro, Fidel, 66
Caucasus, 151
Central European countries and Non-proliferation Regime, 11, 12
Chemical Arms Control, 40, 50, 104
Chemical Warfare, 38
Chemical Warfare, 55
Chemical Weapons, 60, 133, 160
China - Treaties, 151
China, 67, 70, 149, 163, 192, 200,
China-law and Legislation-Nuclear Energy, 98
China-nuclear power plants, 96
China-United States-Commerce, 67
CIS Countries, 177, 216
Clinton Administration, 128, 185
Clinton, U.S. President, 144, 147, 151
Cold War, 45, 68
Collective Security, 119, 167, 197, 198
Committee on Armed Services, 62
Committee on Energy and Commerce, 70, 71
Committee On Foreign Affairs, 72, 73-87
Comprehensive Nuclear Test Ban Treaty (CTBT), 149, 154
Conference on Disarmament, 29
Conference on Global Security, 42
Congress, 71-87
Consequences for U.S. Policy, 37
Constraining Proliferation, 14
Control But Verify, 14

Controlling the atom in the 21st century, 14
Conventional Weapons, 160
Conversion of Industries-Russian Republics, 224
Corporate Corruption-Russia, 215
Corruption in Politics-Russia, 147
Cruise and Ballistic Missiles, 93
CSIS-U.S. -Euratom Senior Policy Panel, 15
Culture AND Security, 15

D

Defense Conversion, 25, 26
Defense Industries-Russia, 224
Defense Industries-U.S., 224
Dept of Defense, 171
Developing Countries-Foreign Relations, 38
Disarmament, 1, 12, 16, 45, 97, 160
Disarmament-Inspection, 114
East Asia-Nuclear Arms Control, 68
Eastern Europe, 145, 199
Economic Assistance, American-Korea (north), 111
Economic Sanctions, 81
Economic Sanctions-Pakistan, 81
Economics, America-Korea, 94
Egypt and the Peaceful uses of Nuclear Energy, 18, 19
El Salvador, 72
Emerging Nuclear Suppliers, 28
Environment and Natural Resources Policy Division, 47
Environmental Monitoring, 19
Ethics, 182
Euratom, 95, 122 172
Euratom-United States, 105
Europe, 105
European Non-Proliferation Policy, 19, 20
Excess Plutonium Disposition alternatives, 39
Explosive Ordinance Disposal, 102

Export Controls - Former Soviet Republics, 54
Export Controls, 20, 41, 52, 71, 74, 87, 92, 163, 214
Export Controls-Europe, 105
Export Controls-Government Policy-United States, 101
Export Controls-India, 67
Export Controls-Japan, 67
Export Controls-Russia, 137
Export Controls-Ukraine, 137
Export Controls-United States, 51, 67, 81, 105, 108, 109
Export Controls-United States-Evaluation, 110

F

Favored Nation Clause-United States, 92
Fighting Proliferation, 21
Fissionable Materials, 157, 165, 191, 193
Fissionable Materials-Security Measures, 171
Food Irradiation, 148
Foreign Relations Korea-United Sates, 111
Foreign Relations- South Asia, 48
Foreign Relations, 163
Foreign Relations-US - Israel, 128
Former Soviet Republics, 53, 72
France- Foreign Relations, 61
France, 200
Future of Arms Control, 138
Future of Nuclear Non-Proliferation Regime, 23
Future War and Counter-proliferation, 59, 60

G

Global Security, 42
Government Policy - India, 51
Government Policy-Europe, 105
Great Britain, 200, 204

Griffith University. Centre for the Study of Australian-Asian Relations, 114

H

Hazardous Substances, 191
Hazards arising out of the peaceful use of nuclear energy, 37
Himalayas, 168
Hong Kong-China-History, 109
Hong Kong-Transfer of Sovereignty from Great Britain, 1997, 109
House of Representatives, 88
How To Remove Residual Threats, 25
How Western European Nuclear Policy is Made, 26
Human Rights-China, 92
Human Rights-Iran, 85
Human Rights-Iraq, 98
Hussein, Saddam, 66

I

Illegal Arms Transfers- Russia (Federation), 109
Illegal Arms Transfers, 33, 94
Illegal Arms Transfers-Iran, 100
Illegal Arms Transfers-United States, 93
India, 130, 142, 145, 148, 150, 167, 168, 213
India-Foreign Relations, 61
India-France Foreign Relations, 61
India-Nuclear Energy, 66, 67
India-Nuclear Weapons, 52, 56
India's Policy and Options , 38
Inspection (Arms Control), 214, 218, 219, 221
Institute of Electrical And Electronics Engineers, 114
Intelligence Service, 95
Intelligence Services- U.S. , 150
International Atomic Energy Agency Evaluation, 196

International Atomic Energy Agency, 58, 59, 65, 141, 152, 165, 171
International Atomic Energy Agency, General Conference (29th: 1985: Vienna, Austria, 107
International Control of Nuclear Power, 122, 135, 219, 221
International Cooperation in Astronautics, 161
International Cooperation, 27, 95
International Economic Policy and Trade, 47, 76
International Law, 176
International Nuclear Agreements, 35
International Nuclear Trade, 27, 28
International Peace Research Institute, 48, 49
International Relations, 120, 161, 165
International Security , 34, 46, 62, 87, 113
International Seminar on Nuclear War and Planetary Emergencies, 28
Iran, 100, 120, 121, 148, 150, 151, 187
Iran-Civil Rights, 85
Iraq, 71, 72, 82, 98, 121, 146, 148, 187
Iraq-United States-Foreign Relations, 99
Ireland, 204
Israel and the non-proliferation Treaty, 28, 29
Israel Affairs, 128
Israel and the Non-proliferation treaty, 28
Israel, 120l, 140, 151, 167l, 183l, 187
Israel-National Security, 120

J

Japan, 68, 131, 153, 170, 181, 211
Japan-North Korea--Foreign Relations, 153
Japan-Nuclear Energy-Law and Legislation, 75
Japan's Nuclear future, 29

K

Kazakhstan, 147, 201, 206
Kernwaffenverbreitung und internationaler Systemwandel, 30, 31
Khomeini, Ruhollah, 66
Korea (North), 89
Korea (North)-Nuclear Weapons, 99
Korea, 77, 89, 170
Korea-Foreign Relations, 63
Korean Peninsula, 178, 203, 211
Korea-Nuclear Weapons, 81
Korea-Politics and Government, 86

L

Lasers, 168
Latin America, 5, 159, 217
Law and Legislation, 73
Law and Legislation-United States-Nuclear Energy, 98
Lee, Teng-Hui President of Rep of China, 133
Libya, 121
Limiting Nuclear Proliferation, 33
Lyndon B. Johnson, 61

M

Managing Conflict in the Nuclear Age, 35
Middle East , 8, 121, 140, 179, 187, 191, 208
Middle East-Arms Control, 120
Military Assistance, American-Pakistan, 96
Military Assistance, China-Iran, 100
Military Assistance, Russian-Iran, 100, 109
Military Policy, 75, 112, 216
Military Policy-Japan, 125
Military Policy-United States, 113, 114
Military Policy-US, 123
Military Relations, 89

Military Relations-Russian Republic, 224
Military research, 93
Military Research-U.S., 222
Military Strategy, 129
Military Surveillance Congresses, 114

N

National Defense University, 60, 173
National Research Council, 54
National Security Archive, 42
National Security, 38, 66, 73, 90, 97
National Security-East Asia, 125
National Security-Middle East, 104
National Security-South Asia, 61
National Security-United States, 54, 77, 78, 84, 93, 103, 158
NATO countries, 200
NATO military forces, 223
New Nuclear Nations, 37, 38
New Threats: Responding To The Proliferation Of Nuclear, Chemical, And Delivery Capabilities In The Third World, 38
NGO, 16
Non-proliferation Agenda, 14
Non-proliferation and Greek Policy, 17
Non-proliferation in a changing world, 38
Non-proliferation in India and Pakistan, 43
Nonproliferation Treaty (NPT), 11, 23, 177, 178
Nonproliferation, 10
North Atlantic Treaty Organization, 122
North Korea and the Bomb, 34
North Korea Arms Control Agreements, 169
North Korea, 79, 80, 119, 120, 123, 133, 149, 154, 155, 156, 158, 161, 163, 174, 175, 195, 198, 200, 202, 206, 209, 218, 221
North Korea-Foreign Relations, 135
North Korea-Nuclear Facilities, 155
North Korea-Nuclear Weapons, 113

North Korea-United States-Foreign Relations, 80
North Pacific, 153
North-East Asia Pacific, 61
Northeast Asia, 130, 170, 195
NPT Regime, 8
Nuclear Anarchy, 5, 6
Nuclear Arms Control, 39, 46, 49, 59, 61, 65, 74, 75, 76, 88, 97, 101, 102, 104
Nuclear Arms Control-Former Soviet Republics, 105
Nuclear Arms Control-Iraq, 98
Nuclear Arms Control-Korea (North), 111
Nuclear Arms Control-Korea, , 94
Nuclear Arms Control-Middle East, 45, 56
Nuclear Arms Control-South Asia, 43, 45, 56
Nuclear Arms Control-United States , 55, 90, 110
Nuclear Arms Control--Verification , 53
Nuclear Black Market, 33
Nuclear Coexistence, 34
Nuclear Crossroads, 5
Nuclear Diplomacy with North Korea, 62, 63
Nuclear Disarmament in Northeast Asia, 26
Nuclear Disarmament, 42, 48, 50
Nuclear Disarmament-East Asia, 44
Nuclear Energy - Law and Legislation, 95
Nuclear Energy- Pakistan, 51
Nuclear Energy Policy-Japan, 166
Nuclear Energy, 1, 47
Nuclear Engineering, 91, 126, 157
Nuclear Exports-Canada, 125
Nuclear Facilities, 65
Nuclear Fuels, 91
Nuclear Fuels-Russia, 207
Nuclear Industry Japan, 63
Nuclear Industry-China, 67
Nuclear Industry-Former Soviet Republics, 40, 41

Nuclear Industry-Government Policy-United States, 105
Nuclear Industry-United States, 106
Nuclear Legislation, 69
Nuclear Nonproliferation treaty NPT, 164
Nuclear Pakistan and Nuclear India, 56
Nuclear Physics, 126
Nuclear power Plants, 134, 166
Nuclear Reactors--Korea (north), 111
Nuclear Research-North Korea, 155
Nuclear Security Measures- U.S., 121
Nuclear Security Measures, 146, 165, 221
Nuclear Security Measures-Russia, 135, 220
Nuclear Terrorism, 94, 219
Nuclear Terrorism-Russia, 215
Nuclear Testing, 49
Nuclear Warfare, 87
Nuclear Weapon Free Zone, 23, 34, 85
Nuclear Weapon Free Zones-Oceania, 79
Nuclear Weapons Industry-Russia (Federation), 109
Nuclear Weapons Tests, 126, 181
Nuclear Weapons Tests-France, 125
Nuclear Weapons-Government Policy-Case Studies, 57
Nuclear Weapons-Government Policy-United States, 104
Nuclear Weapons-Iran, 100
Nuclear Weapons-Iraq, 72, 98

O

Oceania, 74, 85
Oceania-United States-Foreign Relations, 79
Omnibus Export Administration Act Proposed, 156
One Hundred Fifth Congress, First session, 95
One Hundred Fourth Congress-First session-1995, 93
Opaque Nuclear Proliferation, 51
Organized Crime-Russia, 147, 215

Ottawa Verification, 53

P

Pakistan, 8, 48, 65, 80, 81, 82, 96, 120, 130, 142, 148, 167, 168
Pakistan-Foreign Relations, 77
Pakistan-Nuclear Weapons Tests, 150
Pakistan-Nuclear Weapons, 52, 56
Peace Movements, 132
Peace Palace Library, 37
Peaceful Uses of Nuclear Energy, 122, 144, 147, 151
People's Republic of China, 1, 73
Persian Gulf War, 186
Plutonium (OECD) countries, 138
Plutonium and Security, 52
Plutonium debate, 29
Plutonium Industry-Japan, 63
Plutonium, 102, 121, 135, 136, 146, 197, 202, 219, 221, 222
Plutonium-Japan, 125, 165, 199
Plutonium-OECD Countries, 126
Plutonium-Russia, 146
Plutonium-US, 193
Politics and Government-Iraq, 98
Port Controls-Russia, 196
Post Communism, 61
Post-Cold War, 7
Post-cold war nuclear dangers, 174
Present and Future Dangers of Nuclear War, 10
President George Bush (1989-1993), 112
Preventing the spread of nuclear, chemical and biological weapons, 40
Prevention Of Nuclear Power Plant Accidents, 166
Proliferation Treaty Review Conference #rd: 1985: Geneva, Switzerland), 107
Pugwash Conferences on Science and World Affairs, 55, 158
Pulling back from the Nuclear Brink, 56

Q

Qaddafi, Muammar, 66

R

Radiation Safety, 204
Radioactive Substance23, 12
Radioactive Substances-Law and Legislation, 57
Radioactive Waste Disposal, 39, 136, 148, 221, 222
Radioactive Wastes, 126
Reactor Fuel Processing, 136
Reactor Fuel Reprocessing, 52, 95, 138, 199
Reactor Fuels, 207
Reciprocity-China, 92
Regulation of Nuclear Trade, 57
Research and Development Facilities-U.S., 222
Resources of the Future, 59
Review of 1985 YS government nonproliferation activities, 106
Roberts, Guy B., 157
Rotblat, Joseph, 55, 158
Russia, 124, 135, 143, 145, 149, 180, 206, 220
Russian Republic, 224
Russia-U.S., 165

S

Saddam, Hussein, 146
Safety Measures, 65
Saving NTP and Abolishing Nuclear Weapons, 39
Science Policy-U.S., 222
Scientists in Government--Russia (Federation), 115
Scientists, 158
Security in the New World Order, 62
Small States and Security Regimes, 34
South Africa, 63, 162
South Asia, 47, 99, 130, 167, 189, 190, 204, 205
South Asia-United States-Foreign Relations, 100
South Korea, 209
South Pacific Nuclear Free Zone, 74
Southern Pacific Nuclear Free Zone Treaty, 7
Space Sciences, 161
Special Subcommittee on U.S. Pacific Rim Trade, 70
Special Subcommittee on U.S. Trade with China, 70
State-sponsored Terrorism, 66
Strategic Arms Reduction Treaty (START), 149
Strategic Defense initiative, 134
Subcommittee on Arms Control, International Security and Science, 76
Subcommittee on Asian And Pacific Affairs, 74, 77, 79, 81
Subcommittee on Energy Conservation and Power, 71
Subcommittee on International Economic Policy and Trade, 77, 81
Subcommittee On International Operations, 73, 86
Subcommittee on Oversight and Investigation, 71
Subrahmanyam, 46
Supercomputers-United States, 108
Sweden, 13
Symposium on International Safeguards, 65

T

Taiwan, 133
Taiwan, 89
Technical Assistance, American--Former Soviet Republics, 11
Technology and Social Problems, 161
Technology Transfer, 71, 128
Technology Transfer-China, 108
Technology transfer-Russia, 143

Terrorism, 87, 156
The Globalization of defense production, 37
The Indo-Pak nuclear scenario, 51, 52
The Nuclear Asymptote, 35
The Politics of Nuclear nonproliferation, 58
The Spread of Nuclear Weapons: A debate, 58
Thermonuclear War, 49
Trading with the Enemy Ac, 156
Treaties, 153
Treaties, 168
Treaty of Pelindaba, 124
Treaty of Tlatelolco, 159
Treaty on the Non-proliferation of Nuclear Weapons (1968), 84, 86, 103, 123, 131, 152
Tritium-India, 130
Tritium-Pakistan, 130
Twenty years of the Non-proliferation treaty, 25

U

U.S. Dept. of Energy--Rules and Practice, 110
Ukraine, 9, 119, 120, 145, 184, 201, 206, 217
Ukrainian Denuclearization, 119
United Nations Association of the United States of America, 107
United Nations Institute for Disarmament research, 44, 115, 170
United Nations, 52, 69
United Sates military Policy, 113
United States, 68, 69, 70, 71-73
United States Foreign Relations-China, 96
United States Japan Nuclear Agreement, 75
United States Military Policy, 62, 90
United States-China Trade Relations, 91, 92
United States-Defenses, 102

United States-Foreign economic Relations-South Asia, 100
United States-Korea foreign relations, 81
Uranium Enrichment, 91
Uranium Enrichment-Foreign, 218, 219
Uranium, 121
Uranium-Iraq, 127
Uranium-Russia, 207
US Arms Control and Disarmament Agency, 143
US Dept of Energy, 121
US military Policy, 144
Usable Fissile material Storage and excess plutonium, 39

V

Vienna, Austria, 65

W

Waltz, Kenneth Neal, 58
War (International law), 112
Weapons of Mass Destruction- Russia (Federation), 109
Weapons of Mass Destruction, 54, 63, 87, 90, 95, 100, 112, 122, 129, 134, 140, 146, 156, 175, 224
World Politics 1985-1995, 64
World War II, 172
World War III, 50

Y

York Centre for International and Strategic Studies, 54

Author Index

A

Abrams, Herbert L., 190
Aga Khan Sadruddin, Prince, 49
Albright, David, 186
Alves, Dora, 2
Anthony, T.J., 122
Arnett, Eric, 49, 151
Aronson, Shlomo, 4
Arquilla, John, 4, 5, 208

B

Bacevich, A. J., 203
Bailey, Kathleen C., 6, 52, 149, 154, 206
Baker, John C., 6, 137
Bandow, Doug, 194
Barnaby, Frank, 6, 7, 52, 120
Becher, Klaus, 161
Beck, Harald, 7
Beck, Michael, 194
Bee Ronald J., 7, 191
Bellany, Ian, 44
Bennett, Bruce W, 137
Benson, Sumner, 124, 165
Bernauer, Thomas, 7, 8
Bertram, Christoph, 48
Bertsch, Gary K., 67, 194
Bhaskar, C. Uday, 142
Bhatya, Shyam, 8
Bhiaya, Kotera M., 205
Bhola, P.L., 8
Bidwai, Praful, 132
Bitzinger, Richard, 9
Blacker, Colt D., 44
Blank, Stephen J., 204, 209

Blechman, Barry M., 196
Blix, Hans, 9, 164
Boldrick, Michael R., 213
Boller, Scott C., 213
Booker, Malcolm, 10
Bourque, Lyne C., 53
Bowen, Wyn, 124
Brogden, Peter, 206
Bruce, Robert H., 45
Bukharin, Oleg, 196, 207
Bundy, McGeorge, 10
Bunn, George, 10, 11, 153
Bunn, T. Davis, 11
Burns, William F., 143
Bush, George President (1989-1993), 112

C

Cabasso, Jacqueline, 168
Carasales, Julio C., 185
Carpenter, Ted Galen, 222
Carranza, Mario E., 126
Cass, Frank, 15
Cassata, Donna, 146
Cerbiello, Craig, 126
Chari, P.R., 43, 166
Chauviatré, Eric, 12
Chee, Coung-Il, 202
Chellaney, Brahma, 12, 212
Cheney, Glenn Alan, 12
Cho, Kap-Je, 186
Chow, Brian G., 13, 161
Clinton, U.S. President, 144, 147, 151, 162
Cochran, Thomas, 173
Cohen, Avner, 151, 191

Cohen, Stephen Philip, 48
Cole, Paul M., 13, 201
Coll, Steve, 170
Cooper, Mary H., 183
Cozic, Charles P., 44, 45
Crouch, J.D., 180
Cupitt, Richard T., 67, 144

D

Daalder, Ivo H., 182
Davis, Paul K., 208
Davis, Zachary, 54, 163
Davydov, Valery, 15
Dembinski, Matthias, 16
Dewitt, David, 42
Dhanjal, Gursharan S., 16
Dipankar Banerjee, 61
Doherty, Carroll J., 158
Dokos, Thanos P., 17, 18
Domenici, Pete V., 148
Donnelly, Warren H., 18
Doucet, Ian, 192
Dowdy, William, 55, 56
Dregger, Alfred, 195
Duk-Min, Yun, 148
Dunn, Lewis A., 119, 138

E

Elbaradei, Mohamed, 158
Ellis, Jason D., 143
Epstein, William, 173, 201
Evan, William M., 46
Evencoe, Paul R., 197

F

Feaver, Peter D., 123
Feiveson, Harold A., 217
Feldman, Shai, 20
Feldman, Shari, 140
Fischer, David, 21, 22
Fisher, Cathleen S., 196

Fisher, Richard D., Jr., 209, 221
Frankel, Benjamin, 50, 51, 54
Freeman, J. P. G., 22
Fujita, Edmundo, 23

G

Gaffney, Frank, J. Jr., 196
Gallacher, Joseph, 44
Gardner, Gary T., 24
Garnett, Sherman W., 145, 162, 184
Gedda, George, 148
George, Alan, 126
Gerardi, Greg, 192
Gills, Barry K., 181
Gjelstad, Jørn, 48, 49
Goldblat, Jozef, 24, 25
Goldemberg, Jose, 217
Goldring, Natalie J., 223
Gonzalez, Iris M., 197
Goodby, James E., 177
Goozner, Merrill, 168
Gordon, Sandy, 204
Graham, Thomas, Jr., 180, 184
Green, Alex Edward Samuel, 25
Greg, G. Allen, 65, 66

H

Hadjor, Kofi Buenor, 50
Hall, Terry, 203
Halverson, Thomas, 199
Ham, Peter van, 26
Han, Yong-Sup, 26, 170, 192
Hart-Landsberg, Martin, 212
Hawes, John, 166
Hawkins, Helen S., 65, 66
Hersh, Seymour, 215
Hogler, Joe L., 141
Holdren, John P., 219
Hoodbhoy, Pervez, 129
Howlett, Darryl, 194, 203

I

Ikle, Fred C., 132
Imai, Ryukichi, 27, 131
Inbar, Efraim, 179

J

Jalonen, Oili-Pekka, 29
Johnson, Rebecca, 29, 30
Johnson, Stuart E., 175
Jones, Gregory S., 161

K

Kadner, Steven P. , 131
Kahan, Jerome H., 210
Kalinin, Alexander V., 216
Kalinowski, Martin, 129
Kapur, K.D., 30
Karp, Aaron, 50
Kay, David A., 186
Keeny, Spurgeon M., Jr., 149
Kelly, Peter, 30
Kessler, J. Christian, 160
Khripunov, Igor, 144, 194, 211
Kiernan, Vincent, 219
Kim, Chong-gil, 66
Kim, Hakjoon, 191
Kincade, William H., 48, 160
Kirchner, Walter L., 185
Koch, Andrew, 124
Koh, Byung Chul, 188
Kotter, Wolfgang, 31
Krasno, Jean, 212
Krass, Allan, 32
Kristensen, Hans, 139
Kumaraswamy, P.R., 183

L

Lake, Anthony, 178
Lamm, Vanda, 32
Landau, Susan, 158
Larkin, Bruce D., 32
Latter, Richard, 193
Leaver, Richard, 32, 33
Lee Rensselaer W., 33
Lee, Dong-bok, 162
Lee, Steven, 182
Lee, Teng-hui, 89
Leigh-Phippard, Helen, 141
Lennon, Alexander, 207
Lewis, William H., 175
Litwak, Robert S. , 45
Lockwood, Dunbar, 184
Lovins, Hunter L., 50
Lundbo, Sten, 138

M

Mack, Andrew, 120, 130, 136, 169
Manning, Robert A., 224
Mansuorov, Alexandre Y., 178
Marom, Ran, 33, 34
Marples, David R., 40, 41
Martel, William C., 34, 189
Masker, John Scott, 34
Masland, Tom, 207
Mataija, Steven, 53, 54
Maull, Hanns W. , 195
May, Michael M., 174
Mazarr, Michael J. , 34, 35, 139, 155, 163
Medeiros, Evan S., 161
Meise, Gary J. , 139
Merrill, John, 218
Miatello, Angelo, 35
Miller, Marvin, 136
Millot, Marc Dean, 215
Molander, Roger C. , 35, 216
Moodie, Michael, 182
Moore, Mike, 163
Morrison, Phillip, 35, 36, 223
Moshes, Arkady, 217
Mozley, Robert Fred,, 36
Muckerman, Joseph E., 225
Müller, Harald, 26, 36, 41, 65, 164, 183
Mussington, David, 37, 200

N

Nair, Vijai K., 212
Newcomb, Mark E., 198
Niksch, Larry A. , 187, 195
Njølstad, Olav, 48, 49
Norris, Robert S., 200
Nuckolis, John H., 174
Nuller, M. Staius, 37

O

Ogunbanwo, Sola, 124
O'Heffernan, Patrick, 50
O'Neill, Kevin, 186
Ottaway, David B. , 170

P

Paine, Christopher, E., 14, 15
Panofsky, Wolfgang K.H., 220
Paranjpe, Shrikant, 51
Park, Sang Hoon, 209, 213
Pastenrak, Douglas, 179
Paul, T. V., 52
Pendley, William T., 189
Perez, Antonio F. , 188
Perry, Todd, 143
Pilat, Joseph F., 185
Piper, Martyn, 129
Pires, Jeong Hwa, 194
Plotts, James A., 192
Plunk Daryl M., 214
Pollack, Daniel, 190
Potter, William C. , 27, 177
Power, Paul F. , 121
Powers, John R., 225
President George Bush (1989-1993), 112

Q

Quester, George H. , 56, 208, 222
Quinlan, Michael, 142

R

Rathjens, George, 185
Rauf, Tariq, 159
Redick, John R., 185
Reiss, Mitchell, 45, 57,58, 183, 216
Rethinaraj, T.S. , 145
Reynolds, Rosalind R. , 157
Roberts, Guy B., 157
Robinson, Tamara C., 150
Rubinstein, Alvin Z., 218
Rudney, Robert, 52, 122, 145
Ruhle, Michael, 200

S

Sagan, Scott Douglas, 58, 217
Sauer, Tom, 59
Scheinman, Lawrence, 59
Schilling, Walter, 168
Schneider Barry R. , 55, 59, 198
Schwarzbach, David A., 137
Schweitzer, Glen, E., 60
Scott, William B. , 135
Seaborg, Glenn Theodore, 60
Segal, Gerald, 200
Sheffer, Gabriel, 129
Shinrikyo, Aum, 156
Shyam, Babu D.,, 62
Sigal, Leon V. , 62, 63
Simpson, John, 194, 223
Sokolski, Henry, 127, 128, 148, 220
Solomon, Kenneth A., 63
Spector, Leonard, 63, 64, 167
Speier, Richard H. , 161
Starr, Chauncey, 134
Steinberg, Gerald M., 64, 140, 130, 190, 212
Stephen, Blank, 9
Subramanian, R. R. , 64, 65
Swisher, Karin, 44, 45
Szilard, Leo, 65, 66

T

Tanaka, Yoshikazu, 202
Tanter, Raymond, 66
Taylor, Terrence, 66
Tertrais Bruno, 129
Timerbaev, Roland M., 153
Tiwari, H.D. , 66, 67
Treverton, Gregory F., 137
Tripodi, Paolo, 120
Tsipis, Kosta, 223
Turpen, Elizabeth A., 131

U

Utgoff, Victor A. , 208, 222

V

Van Creveld, Martin L., 112
Van Orden, Geoffrey, 124
Vanaik, Achin, 132
Vance, Mary A., 113
Von Hippel, Frank, 136, 177

W

Waina, George, 197
Waltz, Kenneth Neal, 58
Wanner, Barbara, 153
Warren, George, 197
Watman, Kenneth, 182
Wendt, James C., 113, 212
Whitmore, D.C.113
Wilkening, Dean, 113, 114, 182
Will, George F. , 210
Williams, Robert H. , 147
Williamson, Richard L., Jr. , 176
Wilson, Michael, 114
Wilson, Peter A., 216
Wolfsthal, Jon Brook, 167
Wolkomir, Joyce, 132
Wolkomir, Richard, 132
Worsley, Peter, 50

Wrobel, Paulo S., 125, 185

Y

Yamamoto, Takehiko, 67
Young, Marilyn J. , 40, 41
Young, Peter Lewis, 155

Z

Zimmerman, Peter D., 224
Zimmerman, Tim, 179
Zinberg, Dorothy S., 115